D0946629

Other books by Howard W. Eves

ELEMENTARY MATRIX THEORY

FUNCTIONS OF A COMPLEX VARIABLE, VOL. 1 AND 2

FUNDAMENTALS OF GEOMETRY

AN INTRODUCTION TO THE HISTORY OF MATHEMATICS

A SURVEY OF GEOMETRY, VOL. 1 AND 2

IN MATHEMATICAL CIRCLES

INTRODUCTION TO COLLEGE MATHEMATICS,
coauthor with C. V. Newsom

AN INTRODUCTION TO THE FOUNDATIONS AND FUNDAMENTAL CONCEPTS OF MATHEMATICS,
coauthor with C. V. Newsom

THE OTTO DUNKEL MEMORIAL PROBLEM BOOK,
editor with E. P. Starke

Translations

INITIATION TO COMBINATORIAL TOPOLOGY,
by Maurice Fréchet and Ky Fan

INTRODUCTION TO THE GEOMETRY OF COMPLEX NUMBERS,
by Roland Deaux

MATHEMATICAL CIRCLES REVISITED

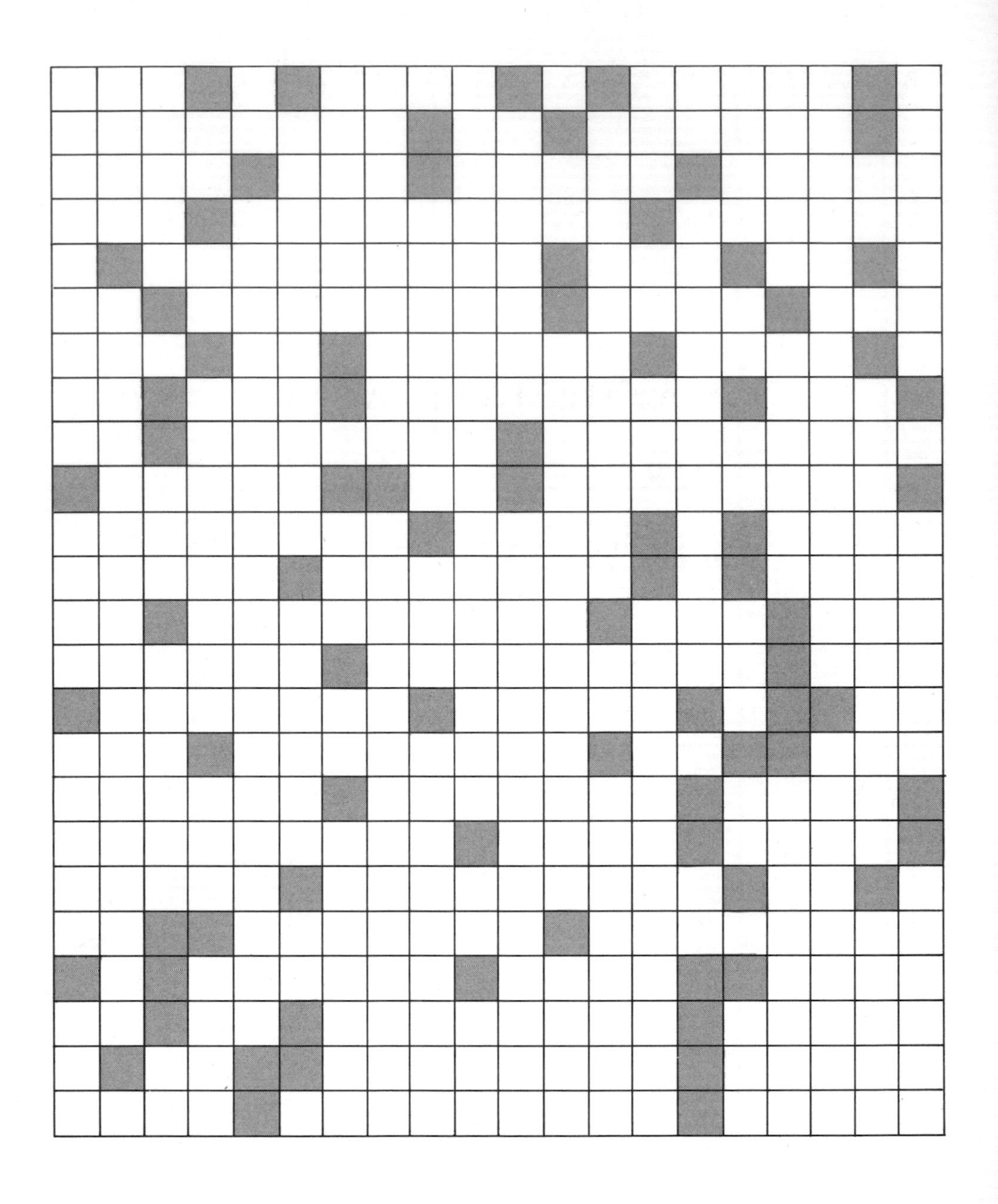

MATHEMATICAL CIRCLES REVISITED

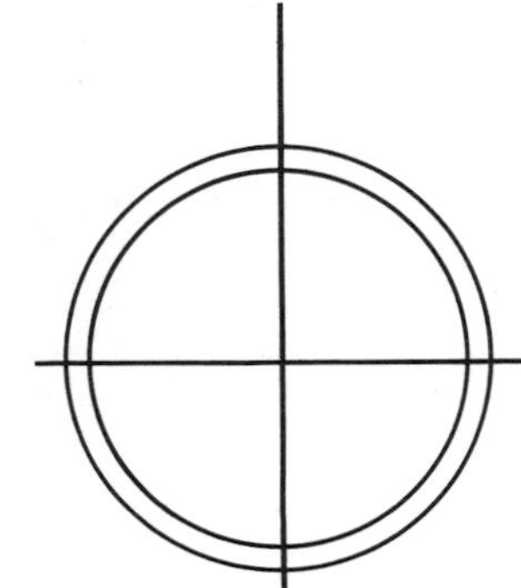

A SECOND COLLECTION OF
MATHEMATICAL STORIES AND ANECDOTES

HOWARD W. EVES

PRINDLE, WEBER & SCHMIDT, INC.

Boston, Massachusetts

FRONTISPIECE: The "random tiling with π" is described in Item 144°.

Library of Congress Catalog Card Number: 71 155300
SBN 87150-121-X

Printed in the United States of America

TO David A. Bradbard
Robert A. Estes
Lincoln T. Fish, jr.
John K. Moulton
Taxia E. Paras
Mary Peabody, Chairman

the other six members of the 1969–1970
Gorham State College mathematics septet
with deep appreciation and warm wishes

PREFACE

The reception given to my little work, *In Mathematical Circles* (Prindle, Weber and Schmidt, Inc., 1969), has been so gratifying, and requests from readers that I make the trip around once again have been so numerous, that I here offer a selection of 360 further mathematical stories and anecdotes. The classification of this second set, in contrast to that of the first set, is more by subject matter than by chronology.

In this selection I have been a bit, but only a *little* bit, bolder concerning anecdotes of living mathematicians. My collection of stories and anecdotes contains many of this type, but I would not for the world want to tread on anyone's feelings, and the task of securing permissions could be prohibitively extensive. So only a few of these contemporary stories appear here, about colleagues of whose good natures I am assured. These are all people I admire, and anything written of them is done so with affection.

The selection is also purposely a mixture of the light and the serious, of the philosophical and the nonsensical, of levity and earnestness, of old and new, of the shallow and the deep. Some people adore puns, others detest them; some lovingly collect boners and bloopers, others are not amused by them; some respond to classroom humor, others do not; some enjoy terse philosophical nuggets, others are bored by them; some find appeal in poignancy, others are distressed by it; and so on. The truth of the matter is that there is room in the teaching profession for all types of stories, and forbearance is asked of any reader possessing a peeve against one type or another. It would seem that perhaps all types should be preserved as part of the rich folklore of the profession.

My sincere thanks go to the colleagues and correspondents who have sent me favorite stories that they said they would like to see in this extended collection. These are all acknowledged by appending the contributors' names to the concerned items. Special thanks go to Alan Wayne, who has long collected bits of mathematical humor, and to George Pólya, master storyteller of our field. Many of the stories about numbers and numerals are adapted from the remarkable and scholarly book, *Number Words and Number Symbols, a Cultural History of Numbers* (M.I.T. Press, 1969), by Karl Menninger. Many of the Bourbaki stories stem from a charming narration by Paul R. Halmos in the *Scientific American*. Also, once again I want to thank *The Mathematics Teacher* for allowing me to adapt some stories that appeared in the Historically Speaking section of that journal, a section which I edited for a number of years. Similar thanks go to *The American Mathematical Monthly* for allowing me to use parts of two of its articles. Several of my personal stories stem from my long association with this excellent collegiate journal. Finally, as in *In Mathematical Circles*, certain historical comments and capsules have been adapted from my book, *An Introduction to the History of Mathematics* (Holt, Rinehart and Winston, third edition, 1969); there the interested reader can find extended historical treatments. Undoubtedly many other acknowledgments should be made. The gradual and largely mental accumulation of some two thousand mathematical stories and anecdotes has occupied a goodly number of years, and the passage of time has lost most of the original sources. I accordingly beg indulgence from anyone who may feel that I have stolen his thunder.

The manuscript for this work was written intermittently during a year spent assisting at Gorham State College, one of the units of the recently formed all-state University of Maine. I certainly owe the administrators and the mathematics staff of Gorham State College deep thanks for making my visit both pleasant and productive. I shall long remember the fine associations I made during this visit.

HOWARD W. EVES

CONTENTS

Contents

QUADRANT TWO

Contents

Classroom Tactics and Antics

Mathematicians and Mathematics

CONTENTS

MILLER TO NEWTON

PEANO TO SWIFT

SYLVESTER TO WHITEHEAD

QUADRANT ONE

From rational digits
to the metric system

NUMBERS AND NUMERALS

Our earliest brush with mathematics—and it occurred when we were the merest of children—concerned itself with numbers and numerals. It seems fitting, therefore, to commence with some stories about these numbers and numerals. Since it would be easy to devote a whole pamphlet to this area alone, we limit ourselves to a couple of dozen stories.

1° *Rational digits.* Mrs. Abdelkri Boujibar, director of the Museum of Morocco, recently publicized an interesting and logical way in which the so-called Arabic numerals 0 through 9 were conceived by a certain Arabian mathematician of over a thousand years ago. He shaped the numerals, as shown in Figure 1, so that each contains the appropriate number of angles. Thus his numeral for "one" contains one angle, that for "two" contains two angles, that for "three" contains three angles, and so on. Zero, of course, was designed so as to contain no angles.

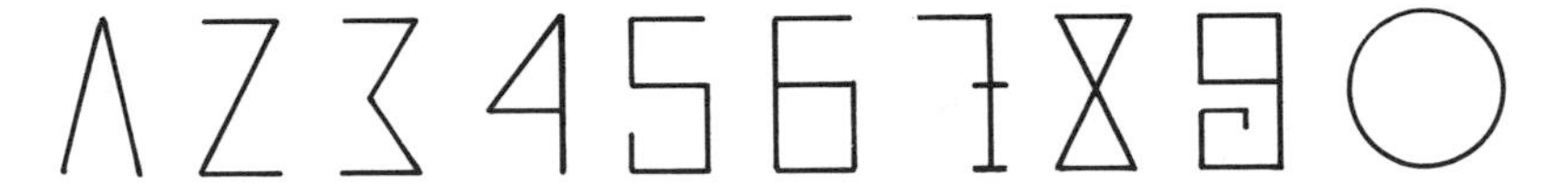

Figure 1

One would like to think that our numerals for the digits actually originated in some such rational way as that described above, but if they did time has obscured the process. The numerals 1, 2, and 3 probably came from cursively writing one, two, and three strokes, in the two latter cases the strokes being horizontal ones. But possible origins of the numerals for the other digits have become confused. The tenth- or eleventh-century Arabic astrologer Aben Ragel suggested the fanciful but unhistorical origins indicated by Figure 2(a), and the eighteenth-century Leipzig mechanic Jacob Leupold claimed that the accountings indicated by Figure 2(b) were widely current in his day. Such imagined explanations arise from the desire of novices to discover the key to a mystery that has eluded experts.

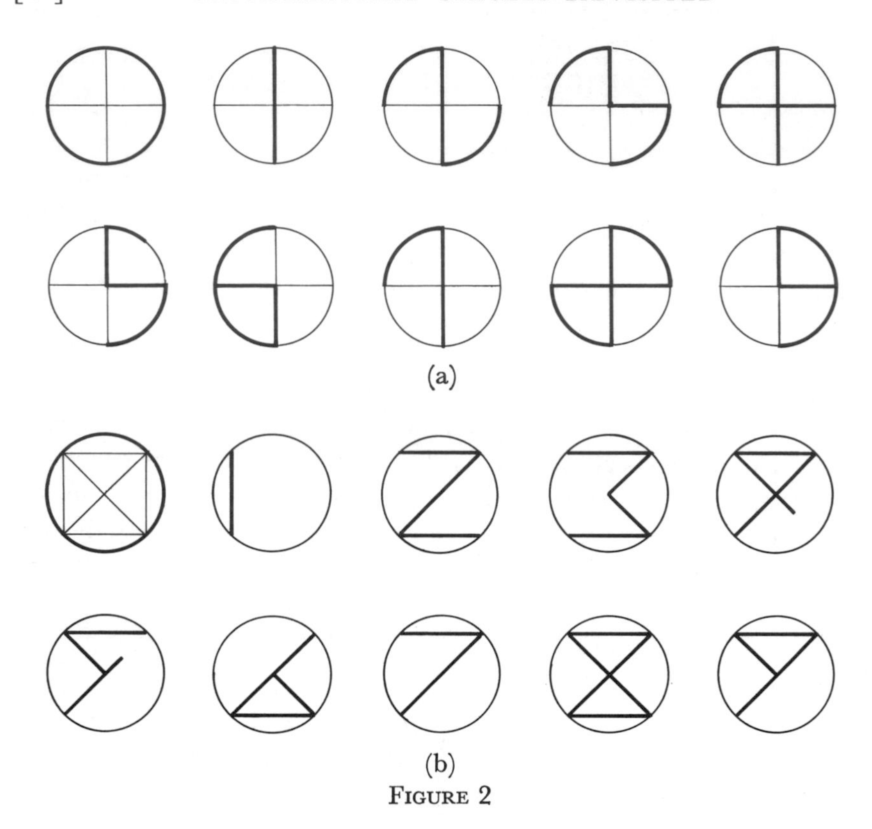

FIGURE 2

2° *Chinese rod, or stick, numerals.* Very interesting is a Chinese numeral system that is probably two thousand or more years old. The system is essentially positional, with base 10. Figure 3(a) shows how the digits 1, 2, 3, 4, 5, 6, 7, 8, 9 are represented when they appear in an odd (units, hundreds, and so forth) position. But when they appear in an even (tens, thousands, and so forth) position, they are represented as shown in Figure 3(b). In this system a circle, ○, was used for zero in the Sung Dynasty (960–1126) and later. Figure 3(c) shows the number 1971 in this system.

A Chinese pictograph for "to compute" is shown in Figure 3(d); it represents some Chinese calculating rods or sticks.

3° *The Babylonian zero.* The earliest positional numeral system that has come down to us is that employed by the ancient Babylonians

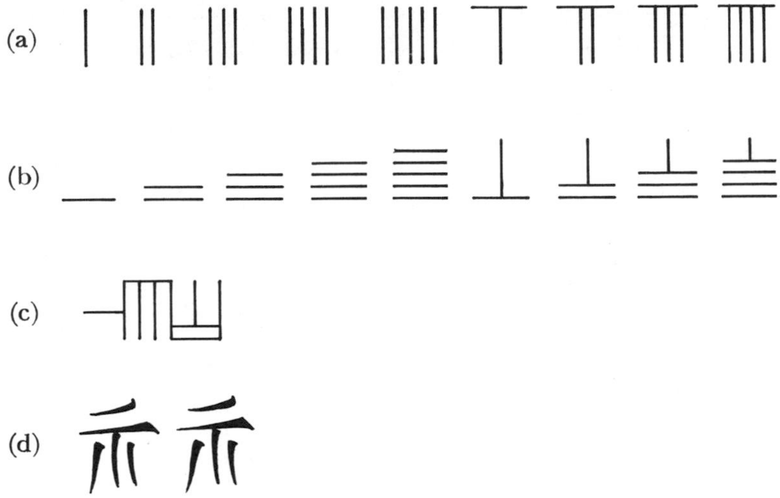

(a)

(b)

(c)

(d)

FIGURE 3

sometime between 3000 and 2000 B.C. It is a sexagesimal system (that is, one based on 60) wherein numbers less than 60 are written by a simple grouping system based on 10. The symbols for "one" and "ten" were

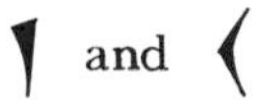

and

respectively. Twenty-five would thus appear as

$$25 = 2(10) + 5 =$$

and 524,551 as

$$524{,}551 = 2(60^3) + 25(60^2) + 42(60) + 31 =$$

This positional system suffered, until after 300 B.C., from a lack of a symbol to stand for any missing power of 60, thus leading to possible misinterpretations of given number expressions. The symbol that was finally introduced consisted of two small slanted wedges. But this symbol was used only to indicate a missing power of the base 60 *within* a number, and not for any missing power of the base 60 occurring at the *end* of a number. The symbol was thus only a partial *zero*, for a true zero serves for missing powers of the base both within and at the end of numbers, as in our 304 and 340. Thus 10,804 would, in the Babylonian, appear as

$$10{,}804 = 3(60^2) + 0(60) + 4 =$$

and 11,040 as

$$11{,}040 = 3(60^2) + 4(60) =$$

rather than as

4° *Papuan body counting.* "Body counting" is the designation of numbers by parts of the human body—such as head, eyes, ears, arms, and so on. Such number sequences are employed by some primitive people. For example, a certain tribe of Papua, in southeastern New Guinea, counts as follows:

1	right little finger	12	nose
2	right ring finger	13	mouth
3	right middle finger	14	left ear
4	right index finger	15	left shoulder
5	right thumb	16	left elbow
6	right wrist	17	left wrist
7	right elbow	18	left thumb
8	right shoulder	19	left index finger
9	right ear	20	left middle finger
10	right eye	21	left ring finger
11	left eye	22	left little finger

One notes the mirror-like repetition in reverse, interrupted by "nose" and "mouth" for 12 and 13.

5° *A curious passage in the Papuan Bible.* Because of a peculiar Papuan system of counting, it was found necessary to translate the Bible passage (John 5:5): "And a certain man was there, which had an infirmity thirty and eight years" into "a man lay ill one man (20), both sides (10), 5 and 3 years."

6° *Three for plural.* Certain primitive societies with only a meager number vocabulary count, "one, two, many." This early conceptual stage of "three" as "many" led "three" to be used in connection with plurals. Thus in Babylonian, the number word *es*, for "three," became the word-ending to convert singular nouns into corresponding plural ones.

But it is the Egyptian hieroglyphs and the Chinese pictographs that most spectacularly perpetuated the idea of three for plural. Thus there is an ancient Egyptian inscription that reads: "To the King thousands were sacrificed and hundreds were offered up." The hieroglyph or ideogram for "thousands" is the symbol for "one thousand" written down three times in a row, and that for "hundreds" is the symbol for "one hundred" written down three times in a row. Elsewhere occur the Egyptian hieroglyphs* shown in Figure 4, in which a

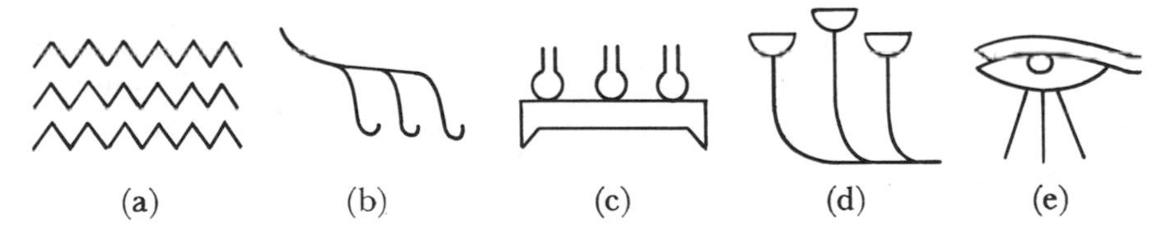

FIGURE 4

repetition in triplicate serves for pluralizing. The ideograms in Figure 4 represent (a) "water" as three (many) waves, (b) "hair" as three

* The hieroglyphs of Figure 4 and the pictographs of Figure 5 are displayed in Karl Menninger's *Number Words and Symbols, a Cultural History of Numbers.*

single hairs, (c) "flood" as heaven with three water jugs, (d) "many plants" as three plants, (e) "weeping" as an eye with three tears. Frequently in Egyptian hieroglyphs, the plural is formed by the ideogram for the singular closely followed by three vertical strokes.

Figure 5 shows five Chinese pictographs using repetitions in triplicate to indicate plurals. Thus Figure 5(a) represents "forest" as

FIGURE 5

three trees, (b) represents "fur" as three hairs, (c) represents "everybody" as three men, (d) represents "endless speaking" as three mouths, (e) represents "gossip" as three women.

Karl Menninger tells of an old Malaccan Sakai's replying, when asked his age, "Sir, I am three years old."

See, also, Item 11° of *In Mathematical Circles*.

7° *Peak-finger days.* Though primitive peoples count on their fingers, only very rarely are the names of the various fingers actually used by them as number words. An exception is the case of the South American Kamayura tribe, who use the word "peak-finger" (middle finger) as their word for "three." Thus "three days" comes out as "peak-finger days."

8° *Number gestures.* It is common among primitive people, and even among sophisticated people, to accompany verbal counting with gestures. For example, in some tribes the word "ten" is frequently accompanied by clapping one hand against the palm of the other, and the word "six" is sometimes accompanied by passing one hand rapidly over the other. Karl Menninger says that certain African tribes can be identified and ethnically classified by observing whether they begin to count on the left hand or the right hand, whether they unfold the fingers or bend them in, or whether they turn the palm toward the body or away from the body.

The Englishman R. Mason has related a charming anecdote about World War II. A Japanese girl was in India, which at the time was at war with Japan. To avoid a possibly embarrassing situation, her friend introduced her as Chinese to an English resident of India. The Englishman was skeptical and asked the girl to count to five on her fingers, which, after some hesitation, she did. Then:

> Mr. Headley burst out delightedly: "There you are! Did you see that? Did you see how she did it? Began with her hand open and bent her fingers in one by one. Did you ever see a Chinese do such a thing? Never! The Chinese count like the English. Begin with the fist closed. She's Japanese!" he cried triumphantly.

9° *The earliest mathematical artifact.* For a long time the oldest extant artifact of some mathematical significance was a royal Egyptian mace, residing in one of the museums at Oxford, believed to date from about 3100 B.C. On the mace are several numbers in the millions and hundred thousands, written in Egyptian hieroglyphs, recording exaggerated results of a successful military campaign.

We now have a considerably earlier artifact, also indicating an involvement with counting. It is a bone tool handle, bearing notches arranged in definite numerical patterns, along with a piece of quartz fixed into a narrow cavity at the head of the handle. Known as the *Ishango bone*, it was found by Jean de Heinzelin a few years ago at Ishango, on the shore of Lake Edward in the Democratic Republic of the Congo, and dates back to the period between 9000 B.C. and 6500 B.C.

Thus it may very well be that the first mathematics on the earth originated, not in Egypt or Mesopotamia, but among the black Africans living south of the Sahara.

10° *Kononto.* Interesting is the word *kononto*, for "nine," of the Mandingo tribe of West Africa. The word literally means "to the one of the belly"—a reference to the nine months of pregnancy.

11° *Number taboos.* [The following is adapted, with permission, from the article, "Black African traditional mathematics," by Claudia

Zaslavsky, which appeared in the Historically Speaking section of *The Mathematics Teacher*, April, 1970, pp. 345–356.]

An interesting phenomenon among certain African tribes is the compounding of the names for seven, eight, or nine; $7 = 6 + 1$, $9 = 8 + 1$, as among the Mandyakos of the western Sudan, while in the Ga language of the Ivory Coast lagoons, $7 = 6 + 1$ and $8 = 6 + 2$. This phenomenon may have been due to a taboo on speaking the names of certain numbers. Seven was a particularly ominous number among the Kongo and Mossi.

The strangest example I have found of a compound name for a number occurs in Umbundu, a language spoken in Portuguese Angola. The name for seven literally means "six—two"! I was fortunate in being able to discuss this matter with a missionary, Mr. Lawrence Henderson, who had spent many years in Angola. He told me that he had it from an Umbundu-speaking person, who in turn had been told by an older person, one who had had no contact with anthropologists, that the original word for seven was subject to a taboo; therefore the word for eight had slipped into its place.

The taboo may be handled in another manner. The speaker merely makes the gesture for the forbidden number, while the listener says the word. In that way the danger is divided between them.

Many African peoples believe it is unlucky to count people, domestic animals, or valuable possessions, for fear that harm may befall them. Among the Kpelle people of Liberia, most arithmetic operations are performed with the aid of piles of stones, and the people become extremely adept in recognizing numbers in this way. In fact, in an experiment requiring estimation of the number of stones in piles of various sizes, Kpelle illiterate adults achieved far better scores than did Yale undergraduate students. They also excelled in estimating the amounts of rice in containers; the activity of measuring rice was an integral part of their lives. However, the Kpelle subjects did poorly when asked about the number of people or houses in their village. Observers of these experiments mentioned the reluctance of the people to count living creatures aloud, but apparently did not realize that fear might lie at the root of the incorrect replies to their questions about the number of people or houses.

Such number taboos are not found only in Africa. The Old Testament relates a taboo on counting people, similar to that of the Kpelle. The ancient Hebrews also had a taboo on the writing of the numerals for fifteen and sixteen. The Hebrew number symbols resembled the Greek, in that the letters of the alphabet represented the numerals. Since the Hebrews used a denary system, fifteen normally would be represented by *yod*, the letter for ten, followed by *heh*, the letter for five. However, this combination, as well as that for ten plus six, would spell the forbidden name of Jehovah. Therefore the combination $9 + 6$ was substituted for $10 + 5$, and $9 + 7$ for $10 + 6$ in written Hebrew. For subsequent numerals, the appropriate digit was added to 10.

And how many buildings in the United States skip "thirteen" in numbering the floors?

12° *Our mixed culture.* Our language is Germanic, our writing is Roman, our numerals are Indian!

13° *The eleventh commandment.* "Thou shalt not divide by zero."

14° *A fair approximation.*

> There was an old man who said, "do
> Tell me *how* I should add two and two?
>   I think more and more
>   That it makes about four—
> But I fear that is almost too few."
>
> ANONYMOUS

15° *A significant truth.* On a sign at the gate of the École de Punaauia in Tahiti appears the equation: "$2 + 2 = 4$."

16° *Fibonacci humor.* A sign in a self-service elevator was observed to read: "Eighth floor button out of order; please push five and three."

17° *The ubiquitous number five.* [The following bit of Pythagoreanism is adapted, with permission, from the note by Professor I. A.

Barnett, of the same title, that appeared in the Historically Speaking section of *The Mathematics Teacher*, April, 1968, pp. 433–435.]

It is of interest to observe how the number five occurs in different parts of mathematics. Below is a list of such occurrences. No doubt others can be added.

1. *Five* points uniquely determine a conic.

2. There are exactly *five* regular polyhedra.

3. The alternating groups of order *five* or less are simple.

4. The general algebraic equation of degree *five* or higher cannot be solved in terms of radicals.

5. All groups of order *five* or less are commutative.

6. *Lamé's theorem.* The number of divisions required to find the greatest common divisor of two numbers is never greater than *five* times the number of digits in the smaller number.

7. There are just *five* complex Euclidean quadratic fields, namely, the fields $a + b\sqrt{m}$, where $m = -1, -2, -3, -7, -11$. (A Euclidean field is a field in which there is a Euclidean algorithm.)

8. There are just *five* known Fermat primes, that is, primes of the form $2^{2^t} + 1$. (The values $t = 0, 1, 2, 3, 4$ yield the Fermat primes 3, 5, 17, 257, 65537.)

9. *Five-color theorem.* Every map on a sphere can be properly colored by using at most *five* different colors. (A map is regarded as properly colored if no two regions having a whole segment of their boundaries in common receive the same color. It is conjectured that four colors will suffice.)

10. The curve

$$ax^3 + bx^2y + cxy^2 + dy^3 = 1,$$

where a, b, c, d are integers, passes through at most *five* lattice points.

11. Every positive integer can be expressed as the sum of at most *five* distinct positive or negative integral cubes. (It is conjectured that four will suffice.)

12. Euler conjectured that, for $n > 1$, at least n distinct positive nth powers are required to sum to a positive nth power. *Five* is the

smallest permissible value of n for which the Euler conjecture is false. (It was recently shown that

$$27^5 + 84^5 + 110^5 + 133^5 = 144^5.$$

It is known that it takes at least four distinct positive fifth powers to yield a positive fifth power, and the above is the smallest instance of such a representation.)

18° *Some more Pythagoreanism?* In studying non-Euclidean geometry, one soon notices the repetition of digits in the dates of many of the famous contributors connected with the early history of the subject. We here list the cardinal dates in the history of non-Euclidean geometry, starting with the European publication of Proclus's *Commentary on Euclid, Book I.* It was this publication of Proclus's work that reawakened interest in Euclid's parallel postulate.

Proclus's *Commentary on Euclid, Book I*, published in Europe in 1533.

J. Wallis, 1616–1703; 1663, the year in which Wallis believed he had "proved" the parallel postulate.

G. Saccheri, 1667–1733; 1733, the year in which Saccheri's work on the parallel postulate was published.

J. H. Lambert, 1728–1777; 1766, the year in which Lambert wrote his unfinished work on the parallel postulate, exactly 33 years after the appearance of Saccheri's work; 1788, the year friends of Lambert put his work through the press, exactly 11 years after Lambert had died.

A. M. Legendre, 1752–1833.

C. F. Gauss, 1777–1855.

J. Bolyai, 1802–1860; 1823, 1829, 1832, important dates connected with Bolyai's work on the parallel postulate.

N. I. Lobachevsky, 1793–1856; 1829–30, 1840, important dates connected with Lobachevsky's work on the parallel postulate.

Prior to Bolyai and Lobachevsky, repeated digits appear in the above dates; they cease to appear when we reach Bolyai. Surely there must be some Pythagorean meaning here. Perhaps it is this: The workers prior to Bolyai felt that the parallel postulate is deducible from Euclid's other postulates, and that any denial of the parallel postulate must involve contradictions. At least this is true of all the workers prior to Bolyai except Gauss. But Gauss, though he did believe the denial of Euclid's parallel postulate could not lead to any contradictions, never had the courage to say so publicly. The repeated-digit spell is broken when we come to Bolyai and Lobachevsky because each of these men in no uncertain terms, and for the first time in history, publicly expressed a belief that the parallel postulate is independent of Euclid's other postulates, and therefore cannot be deduced from them.

19° *Number mysticism.* Charles Sanders Peirce (1839–1914), son of the distinguished Harvard mathematician, Benjamin Peirce, and himself an early contributor to the field of symbolic logic, took from a biographical dictionary of poets the names and ages of death of the first five poets listed. They were

Aagard,	died at 48
Abeille,	died at 76
Abulola,	died at 84
Abunowas,	died at 48
Accords,	died at 45

From the five ages he noted the following common characteristics:

1. The difference of the digits in each age leaves a remainder of *one* when divided by *three*.
2. The first digit in each age, raised to the power indicated by the second digit, leaves a remainder of *one* when divided by *three*.
3. The sum of the prime factors of each age, including *one* as a prime factor, is divisible by *three*.

20° *Giving credit to man. If* "Number rules the universe" as Pythagoras asserted, Number is merely our delegate to the throne, for we rule Number.—E. T. BELL

21° *Giving credit to diety.*

1. God ever geometrizes.—PLATO
2. God ever arithmetizes.—C. G. J. JACOBI

3. The Great Architect of the Universe now begins to appear as a pure mathematician.—J. H. JEANS

4. God made the integers, all the rest is the work of man.
LEOPOLD KRONECKER

5. Nature's great book is written in mathematical symbols.
GALILEO

22° *The Kensington stone mystery.* [The following little detective story, as told by Dr. D. J. Struik of the Massachusetts Institute of Technology, is here repeated, with permission, from the Historically Speaking section of *The Mathematics Teacher*, March, 1964, pp. 166–168.]

One day in 1898, we are told, the Swedish immigrant farmer Olof Ohman, clearing a timbered section of his land near Kensington, Minnesota, found a stone under the roots of a tree. It was soon discovered that this stone was inscribed with strange symbols which experts eventually declared to be runes—characters from an ancient letter script used in Nordic countries for more than a thousand years. These runes could be deciphered, and they conveyed the rather astonishing information that in the year 1362 a group of eight Goths and twenty-two Norsemen (Swedes and Norwegians), on an exploration from Vinland to the West, had suffered an attack in which ten men were killed. The attack occurred on an island fourteen days' journey from their ships.

This indeed was a pre-Columbian document with a vengeance. As early as the fourteenth century there were Scandinavians in Minnesota! The Kensington stone, under the able sponsorship of Mr. H. R. Holand, a Norwegian-American writer, made many headlines, kept many pens busy, and received semiofficial recognition through a year's exhibit at the Smithsonian Institution in Washington, D. C. (1948–49). A gigantic reproduction in "Runestone Memorial Park" near Alexandria, Minnesota, was unveiled in 1951. Many experts and semiexperts have expressed their conviction that the stone is indeed a genuine record from a bygone age.

However, there have always been plenty of experts and semiexperts who were skeptics or even downright unbelievers. Among them we find Professor Eric Wahlgren, specialist in Scandinavian languages, who has written a book exposing the inscription as a hoax. Since many of the arguments, pro and con, deal with the circumstances under which the stone was found and with the interpretation of the runes, the mathematician can only stand by, listen, and be entertained. But there is one argument on which he can have a professional opinion, and this is weighty enough to make him a partisan in the amusing controversy.

Six numbers appear in the inscription—2, 8, 10, 14, 22, and 1362. The 2 and the 8 offer no difficulty, since they seem to have been written in recognized runes. The 10 is represented by an unusual rune, something like a circle with a 1 through it, and we may leave it to the runologists to debate its acceptability. But the 14, 22, and, above all, the 1362, are of a different nature. These numbers are written not only in the decimal system (nothing exceptional), but in the decimal place-value system with runes as digits! Now the decimal system with place value was, to a certain extent, known in the Scandinavia of those days. It had gradually penetrated into Europe from the Islamic countries and can be found, with many clever applications, in Leonardo of Pisa's *Liber Abaci* (1202). Later it was taught in several much simpler texts often called "Algorithmus," of which that written in about 1250 by Johannes de Sacrobosco (probably an Englishman who taught at the University of Paris) was widely studied for centuries, also in Scandinavia. We know of an Icelandic translation of Sacrobosco's book from the fourteenth century, which may be the same as the "Algorismus" contained in the *Hauksbok*, a manuscript of about 1320 preserved in Copenhagen. All those texts, however, write the numbers with the special digits taken (with variations) from the Arabic texts, and these so-called Arabic, or Hindu-Arabic, numerals already look very much like the numerals 0, 1, 2, . . ., which we use. Hardly anywhere has a serious attempt been made to use other digits taken from previous systems of numeration, and that for the simple reason that it would have been a clumsy step backward, liable to all kinds of ambiguities.

An exception to this rule is found on the Kensington stone. Here the carver wrote 14, 22, 1362 with the runes for 1, 2, 3, 4, and 6. This is as if a medieval scribe wrote 14 as I IV with Roman, or 22 as ββ with Greek, numerals. In those days scribes did write numbers in Roman or Greek numerals, but in the way appropriate to that system, as κβ for 22, or MCCCLXII for 1362. There are also examples of runic expressions for numbers, as those given for the numbers 1 to 19 by Ole Worms, but again, they are not in the place-value system.

We have, therefore, an alternative. Either the stone is genuine, and the group of unfortunate Vikings, stranded thousands of miles from home on an island in a bleak wilderness, had with them a cultured scribe who had studied Sacrobosco or a similar author, so that he knew the principle of place value, but who in his misery had forgotten the Arabic symbols and substituted the runes he still remembered, despite the fact that there were customary runic and also nonrunic, non-Arabic, ways of expressing numbers such as 14, 22, etc., or the stone is a hoax, and the runic way of writing 14, 22, etc., in the place-value system is exactly what a Scandinavian wit of 1898 in possession of a book with runes would have done, since he could not have been expected to know the specialized literature on the decimal positional system, of which there certainly was very little in American libraries, not to mention those of pioneer Minnesota.

We do not claim that there were no Goths or Norsemen in Minnesota during the fourteenth century. Mr. Holand and his supporters are welcome to all of them. The only thing we claim is that the inscription on the Kensington stone as an argument to prove their presence in the Mississippi basin during the fourteenth century is not really very convincing.

23° *The gelosian algorithm.* The arithmetics of the fifteenth and sixteenth centuries contain descriptions of algorithms for the fundamental operations. Of the many schemes devised for performing a long multiplication, the so-called *gelosia*, or *grating*, method was perhaps the most popular. The method, which is illustrated in Figure 6 by the multiplication of 9876 and 6789 to yield 67,048,164, is very old. It was probably first developed in India, for it appears in a commentary on the *Lilāvati* of Bhāskara (1114–ca. 1185) and in other Hindu works. From India it made its way into Chinese, Arabian, and Persian literature. It was long a favorite method among the Arabs, from whom it passed over to the Western Europeans. Because of its simplicity in application, it could well be that the method might still be in use but for the inconvenience of printing, or even drawing, the needed net of lines. The pattern resembles the grating, or lattice, used in some windows. These were known as *gelosia*, eventually becoming *jalousie* (meaning "blind," in French).

Note that the additions are performed diagonally and, because of the way each cell is divided into two by a diagonal, no carrying over is required in the partial multiplications.

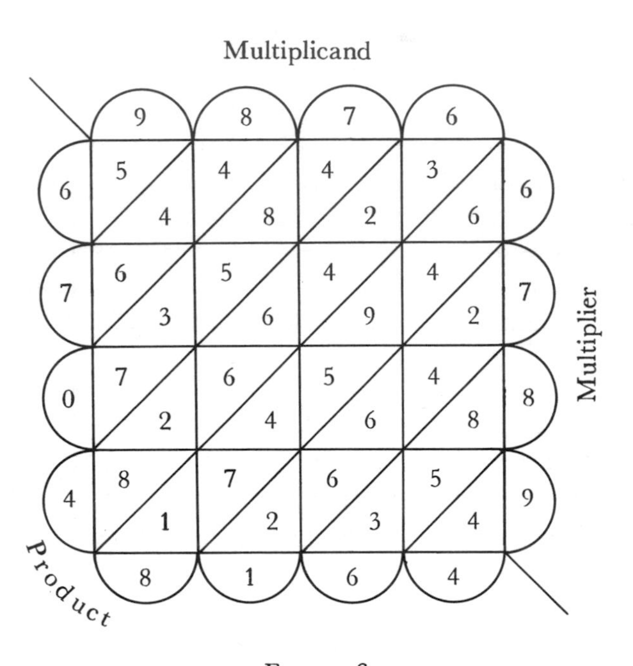

FIGURE 6

24° *The galley algorithm.* By far the most common algorithm for long division in use before 1500 was the so-called *galley*, or *scratch*, method, which in all likelihood was of Hindu origin. To clarify the method, consider the following steps in the division of 9413 by 37.

1. Write the divisor, 37, below the dividend as shown. Obtain the first quotient digit, 2, in the usual manner, and write it to the right of the dividend.

```
9413 | 2
37
```

2. Think: 2 × 3 = 6, 9 − 6 = 3. Scratch 9 and 3 and write 3 above the 9. Think: 2 × 7 = 14, 34 − 14 = 20. Scratch 7, 3, 4 and write 2 above the 3 and 0 above the 4.

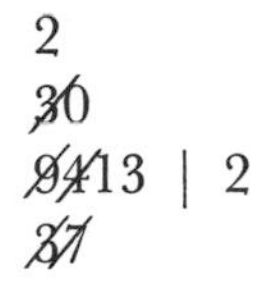

```
 2
 30
 9413 | 2
 37
```

3. Write the divisor, 37, one place to the right, diagonally. The resultant dividend after Step 2 is 2013. Obtain the next quotient digit, 5. Think: 5 × 3 = 15, 20 − 15 = 5. Scratch 3, 2, 0 and write 5 above the 0. Think: 5 × 7 = 35, 51 − 35 = 16. Scratch 7, 5, 1 and write 1 above the 5 and 6 above the 1.

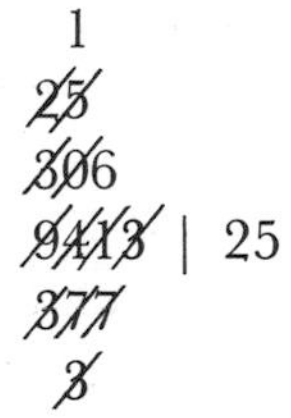

```
   1
  25
 306
 9413 | 25
 377
  3
```

4. Write the divisor, 37, one more place to the right, diagonally. The resultant dividend after Step 3 is 163. Obtain the next quotient digit, 4. Think: 4 × 3 = 12, 16 − 12 = 4. Scratch 3, 1, 6 and write 4 above the 6. Think: 4 × 7 = 28, 43 − 28 = 15. Scratch 7, 4, 3 and write 1 above the 4 and 5 above the 3.

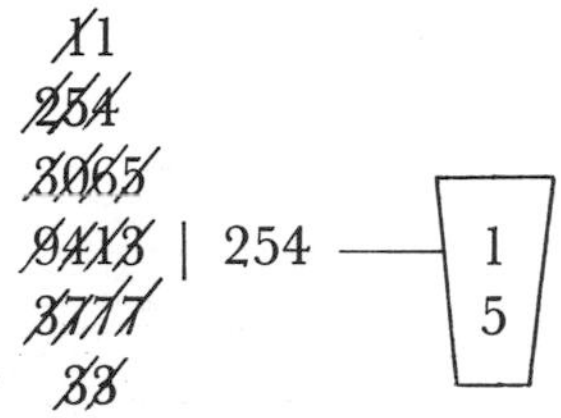

```
  11
 254
 3065
 9413 | 254 ——[1]
 3777         [5]
  33
```

5. The quotient is 254, with remainder 15.

After a little practice, the galley method is not nearly as difficult as it at first appears. Its popularity was due to the ease with which it can be used on a sand abacus or slate, where the scratching is actually a simple erasing followed by a possible replacement. The name "galley" referred to a boat, which the outline of the finished problem was thought to resemble. The resemblance follows either by viewing the work from the bottom of the page, when the quotient appears as a bowsprit, or by viewing the work from the left side of the page, when the quotient appears as a mast. In this second viewpoint, the remainder was frequently written (as indicated above) to look like a flag on the top of the mast.

BIG NUMBERS

CHILDREN, lovers, and governments indulge in large numbers—children in their boasting, lovers in measuring the magnitude of their affections, and governments in taxation and national debts.

25° *The Tower of Hanoi.* The game known as the Tower of Hanoi was brought out in 1883 by M. Claus, Mandarin of the College of Li-Sou-Stian—anagram for M. François Edouard Anatole Lucas (1842–1891), then a professor of mathematics at the Lycée Saint-Louis. The game consists of a board with three vertical pegs fastened to it. At the start of the game there are a number of washer-like disks of different radii arranged on one of the pegs so that the largest disk is on the bottom, the next largest on that one, and so on, up to the smallest disk on top (see Figure 7). The problem is to transfer the set of disks

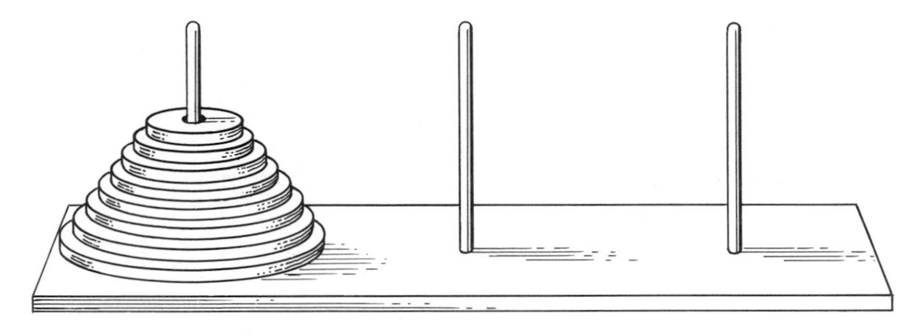

FIGURE 7

to one of the other pegs, moving only one disk at a time and never permitting any disk to rest on a smaller one; all three pegs may be used. The mathematical theory of the game reveals that if there are n disks then a minimum of $2^n - 1$ transfers are required.

The inventor of the game told the following pretty conceit concerning the game's origin. When Mandarin Claus was traveling in connection with research on the works of Fer-Fer-Tam-Tam [Fermat], he came upon the great temple at Benares. Beneath the dome of this temple is a brass plate that marks the center of the world, and there are three diamond needles each a cubit high fixed to the plate. On one of these needles God placed, at the Creation, a stack of sixty-four disks of pure gold, decreasing in size from the bottom of the stack up. This stack of disks is the Tower of Brahma, and day and night without rest, the priests of the temple transfer the disks from one diamond needle to another according to the immutable laws of Brahma, which require that the priests move only one disk at a time and never place any disk on a needle already holding a smaller disk. When all sixty-four disks shall in this way finally be transferred from the needle on which God originally placed them to one of the other needles, the tower, the temple, and the priests will crumble into dust, and with a thunderclap the world will vanish.

If the industrious Brahmin priests can make one transfer every second, with never a mistake and working without pause day and night, they will require

$$2^{64} - 1 = 18{,}446{,}744{,}073{,}709{,}551{,}615$$

seconds, or more than 584 billion years!

26° *The number* $\mathbf{2^{64} - 1}$ *again.* There is an old tale to the effect that the Indian King Shirhân granted the Grand Vizier Sissa Ben Dahir a boon for having invented the engaging and challenging game of chess. Since chess is played on a board of sixty-four squares, Sissa addressed the king: "Majesty, give me a grain of wheat for the first square of the board, two grains for the second square of the board, four grains for the third square, eight grains for the fourth square, and so on for all the sixty-four squares of the board." "Is that all you wish, you fool?" exclaimed the astonished king.

A high school mathematics student familiar with the summing of geometric progressions can easily show that Sissa demanded $2^{64} - 1$ grains of wheat, enough wheat, it has been calculated, to cover the entire surface of the earth to a depth of about one inch!

27° *Ancestors.* Assume that each person has two parents, four grandparents, eight great-grandparents, and so on. Allowing no incestuous combinations, it follows that everyone has at least 2^{64}, or somewhat less than eighteen and a half quintillion, ancestors since the beginning of the Christian era—which was just about sixty-four generations ago.

28° *Incredible.* Suppose that a large sheet of paper one one-thousandth of an inch thick is torn in half and the two pieces are put together, one on top of the other. These are then torn in half, and the four pieces put together in a pile. If this process of tearing in half and piling is done fifty times, how high will the final pile of paper be?

Upon being asked this question, many will say the pile will be several feet high, others that it will be approximately a mile high, and still others, throwing all discretion to the winds, will say that it will be something like 100 miles high. Actually, the final pile will be over seventeen million miles high!

29° *Googol and googolplex.* Some years ago Professor Edward Kasner of Columbia University was concerned with such estimates as:

1. the boiling point of iron is 5.4×10^3 degrees Fahrenheit;
2. the temperature at the center of an atomic bomb explosion is 2×10^8 degrees Fahrenheit;
3. the total number of bridge hands is 6.35×10^{11};
4. the total number of words spoken since the beginning of the world is about 10^{16};
5. the total number of printed words since the Gutenberg Bible appeared is somewhat larger than 10^{16};
6. the age of the earth is set at about 3350 million years, or about 10^{17} seconds;
7. the half-life of uranium 238 is 1.42×10^{17} seconds;
8. the number of grains of sand on the beach at Coney Island, New York, is about 10^{20};
9. the total age of the expanding universe is probably less than 2000 million million years, or some 10^{22} seconds;
10. the mass of the earth is about 1.2×10^{25} pounds;
11. the number of atoms of oxygen in an average thimble is perhaps about 10^{27};
12. the diameter of the universe, as assigned by relativity theory, is about 10^{29} centimeters;

13. the number of snow crystals necessary to form the ice age would be about 10^{30};

14. the total number of ways of arranging 52 cards is of the order 8×10^{67};

15. the total number of electrons in the universe is, by an estimate made by Eddington, about 10^{79}.

Professor Kasner found it convenient to have a name for the number 10^{100}, which considerably exceeds any of the above estimated numbers, and he appealed to his nine-year-old nephew for a suitable name. The child suggested "googol," and forthwith the immense number 10^{100} was called a *googol.* Later, the ever-so-much larger number 10^{googol} was christened a *googolplex.*

30° *Eddington's number.* Sir Arthur Eddington claimed, on the basis of physical theory, that there are *exactly*

$$17 \times 2^{259}$$

protons in the universe. That is, he claimed that in the universe there are

$$15{,}747{,}724{,}136{,}275{,}002{,}577{,}605{,}653{,}961{,}181{,}555{,}468{,}044{,}-$$
$$717{,}914{,}527{,}116{,}709{,}336{,}231{,}425{,}076{,}185{,}631{,}031{,}276$$

protons—accurate to the last digit.

31° *An amplification factor.* In a telephone call from San Francisco to London, the original energy of the human voice must be amplified time after time in the process of transmission to a listening ear in London. The over-all amplification factor of this process has been said to be about 10^{256}.

32° *Skewes' number.* A veritable giant among giant numbers is *Skewes' number,* which dwarfs even a googolplex. It is named after the English mathematician Skewes, and it arose in connection with a study of the distribution of primes. It is the number

$$10^{10^{10^{34}}}$$

The total number of possible moves in a game of chess is of the order

$$10^{10^{50}}.$$

Professor G. H. Hardy once pointed out that if we consider the universe as a great three-dimensional chessboard in which the protons are the chessmen, and if we agree to call the interchange of the positions of any two protons a "move" in this cosmic game, then the total number of possible moves is, of all strange coincidences, Skewes' number.

Dr. Warren Weaver, to shock readers into some sort of realization of just the number of digits in Skewes' number, should that number be written out in full, once described the following. Let us think, he said, of using tiny type with which we could print a million microscopic digits per inch. Suppose we could print lines of these digits which would stretch across the diameter of the whole universe (about 4×10^{28} inches). Suppose that from the beginning of the universe (about 2000 million million years ago) a staff of two billion persons had worked day and night printing out such lines of digits across the universe, and that each person had printed at the rate of a million such lines every second of all this time. Then, through all this absolutely unimaginable effort, only about 10^{72} digits could be printed. This could hardly be considered even an effective start in printing out Skewes' number in full.

33° *Perhaps the most remarkable numerical statement in English literature.* Edward Gibbon's statement: "A thousand swords were plunged at once into the bosom of the unfortunate Probus."

34° *A seventeenth-century pun by Richard Lovelace.*

Why are *wise* few, *fools* numerous in the excess?
'Cause, wanting *number*, they are *numberlesse*.

35° *The square root of infinity.*

There was a young man from Trinity,
Who solved the square root of infinity.

While counting the digits,
He was seized by the figets,
Dropped science, and took up divinity.
Anonymous

36° *The mathematicians' love-knot.* Sir Arthur Eddington, in *New Pathways in Science* (1935), says: "That queer quantity 'infinity' is the very mischief, and no rational physicist should have anything to do with it. Perhaps that is why mathematicians represent it by a sign like a love-knot."

PI

One of the most famous of all numbers is that universally designated today by the lower case Greek letter π; it represents, among other things, the ratio of the circumference to the diameter of a circle. It has enjoyed a long and interesting history, and over the years it has received ever better approximations.

37° *π in Biblical times.* In the ancient Orient the value of π was frequently taken very roughly as 3. Thus in the Old Testament, in II Chronicles 4:2 (and similarly in I Kings 7:28), we read: "Also he made a molten sea of ten cubits from brim to brim, round in compass and five cubits the height thereof, and a line of thirty cubits did compass it about." This tells us that the Hebrews of the time approximated the ratio of the circumference to the diameter of a circle by 3. This is equivalent to approximating the circumference of a circle by the perimeter of an inscribed regular hexagon. From this rough approximation of 3, the value of π has been determined more and more accurately, reaching, in 1967, the fantastic accuracy of a half-million decimal places.

38° *An ancient Egyptian estimation of π.* In the Rhind Papyrus, an ancient Egyptian text dating from about 1650 B.C. and composed of eighty-five mathematical problems, Problems 41, 42, 43, and 48 involve finding the area of a circle. In each of these problems the area of a circle is taken equal to the square on (8/9)ths of the circle's diameter.

It is not known how this formula for "squaring a circle" was arrived at, but a little geometrical diagram (here reproduced in Figure 8(a)) accompanying Problem 48 offers a possible clue. It may be that the diagram represents a square with its sides trisected and then the four corners cut off. If such a figure is carefully drawn (see Figure 8(b)), it

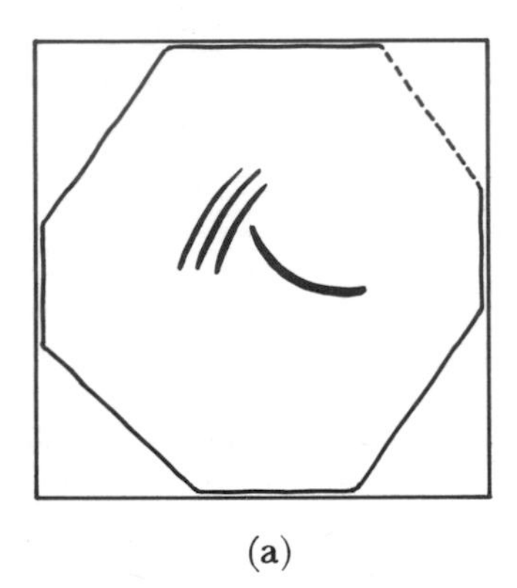

(a)

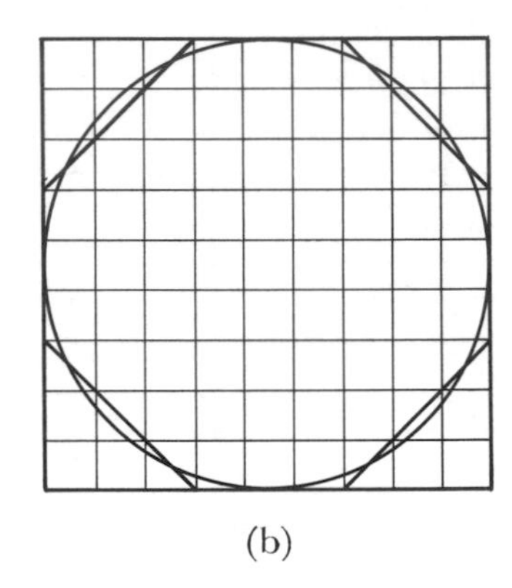

(b)

FIGURE 8

looks by eye as though the area of the circle inscribed in the square is pretty well approximated by the octagon-shaped figure. If the diameter of the circle, and hence the side of the square, be taken as 9, the octagon will have an area equal to

$$81 - 4(9/2) = 63,$$

and the side of a square of the same area as the circle would be approximately $\sqrt{63}$, which, in turn, is approximately $\sqrt{64}$ or 8. That is, the side of the equivalent square is about (8/9)ths of the diameter of the given circle.

If we assume the above Egyptian formula for "squaring a circle" is correct, that is, that

$$\pi d^2/4 = 64d^2/81,$$

we arrive at

$$\pi = 256/81 = (4/3)^4 \doteq 3.1605,$$

which is not too bad an estimation of π for those remote times.

39° *An interesting rational approximation of* π. About the year

480, the early Chinese worker in mechanics, Tsu Ch'ung-chih, gave the interesting rational approximation $355/113 = 3.1415929\cdots$ of π, which is correct to six decimal places and, curiously, involves only the first three odd numbers, each twice. About 1585, Adriaen Anthoniszoon rediscovered the ancient Chinese ratio. This was apparently a lucky accident since all he showed was that

$$377/120 > \pi > 333/106.$$

He then averaged the numerators and the denominators to obtain the "exact" value of π. There is evidence that Valentin Otho, a pupil of the early table maker Rhaeticus, may have introduced this ratio for π into the Western world at the slightly earlier date of 1573.

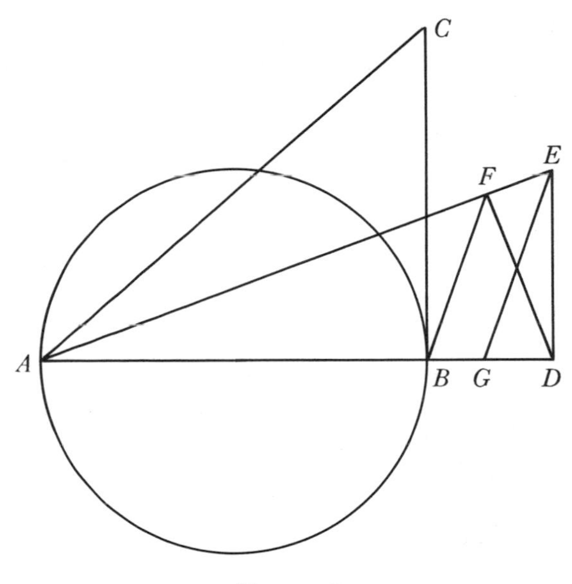

FIGURE 9

In 1849, de Gelder used the ratio 355/113 to obtain a close approximate Euclidean solution of the problem of rectifying a given circle. Let (see Figure 9) $AB = 1$ be a diameter of the given circle. Draw $BC = \frac{7}{8}$, perpendicular to AB at B. Mark off $AD = AC$ on AB produced. Draw $DE = \frac{1}{2}$, perpendicular to AD at D, and let F be the foot of the perpendicular from D on AE. Draw EG parallel to FB to cut BD in G. The reader may now care to show that

$$GB/BA = EF/FA = (DE)^2/(DA)^2 = (DE)^2/[(BA)^2 + (BC)^2],$$

whence $GB = 4^2(7^2 + 8^2) = 16/113 = 0.1415929\cdots$, the decimal part of π correct to six decimal places. The circumference of the circle would then be given very closely by three diameters plus the segment GB.

40° *The Ludolphine number.* Ludolph van Ceulen (1540–1610) of Germany computed π to thirty-five decimal places by the classical method of inscribed and circumscribed regular polygons, using polygons having 2^{62} sides. He spent a large part of his life on this task and his achievement was considered so extraordinary that the number was engraved on his tombstone, and to this day is sometimes referred to in Germany as "the Ludolphine number." Recent attempts to find the tombstone have been unsuccessful; it is probably no longer in existence.

41° *Mnemonics for* π. Among the curiosities connected with π are various mnemonics that have been devised for the purpose of recalling π to a large number of decimal places. The following, by A. C. Orr, appeared in the *Literary Digest* in 1906. One has merely to replace each word by the number of letters it contains to obtain π correct to thirty decimal places.

Now I, even I, would celebrate
In rhymes inapt, the great
Immortal Syracusan, rivaled nevermore,
Who in his wondrous lore,
Passed on before,
Left men his guidance how to circles mensurate.

A few years later, in 1914, the following similar mnemonic appeared in the *Scientific American:*

See, I have a rhyme assisting my feeble brain, its tasks ofttimes resisting.

Two other popular mnemonics are:

How I want a drink, alcoholic of course, after the heavy lectures involving quantum mechanics.

May I have a large container of coffee?

42° *A brief chronology of the calculation of* π *by infinite series.*

1691. The Scottish mathematician James Gregory obtained the infinite series

$$\arctan x = x - x^3/3 + x^5/5 - x^7/7 + \cdots, \qquad (-1 \leqq x \leqq 1).$$

1699. Abraham Sharp found 71 correct decimal places by using Gregory's series with $x = \sqrt{(1/3)}$.

1706. John Machin obtained 100 decimal places by using Gregory's series in connection with the relation

$$\pi/4 = 4 \arctan(1/5) - \arctan(1/239).$$

1719. The French mathematician De Lagny obtained 112 correct places by using Gregory's series with $x = \sqrt{(1/3)}$.

1841. William Rutherford of England calculated π to 208 places, of which 152 were later found to be correct, by using Gregory's series in connection with the relation

$$\pi/4 = 4 \arctan(1/5) - \arctan(1/70) + \arctan(1/99).$$

1844. Zacharias Dase, the lightning calculator, found π correct to 200 places using Gregory's series in connection with the relation

$$\pi/4 = \arctan(1/2) + \arctan(1/5) + \arctan(1/8).$$

1853. Rutherford returned to the problem and obtained 400 correct decimal places.

1873. William Shanks of England, using Machin's formula, computed π to 707 places. For a long time this remained the most fabulous piece of calculation ever performed.

1948. In 1946, D. F. Ferguson of England discovered errors, starting with the 528th place, in Shanks' value of π, and in January 1947 he gave a corrected value to 710 places. In the same month J. W. Wrench, Jr., of America, published an 808-place value of π, but Ferguson soon found an error in the 723rd place. In January 1948, Ferguson and Wrench jointly published the corrected and checked value of π to 808 places. Wrench used Machin's formula, whereas Ferguson used the formula

$$\pi/4 = 3 \arctan(1/4) + \arctan(1/20) + \arctan(1/1985).$$

1949. The electronic computer, the ENIAC, at the Army Ballistic Research Laboratories in Aberdeen, Maryland, calculated π to 2037 decimal places, taking 70 hours of machine time.

1954. Nicholson and Jeenel, using NORC, calculated π to 3089 places in 13 minutes.

1958. Felton, in England, using a Ferranti PEGASUS computer, calculated π to 10,000 decimal places in 33 hours.

1958. François Genuys, in Paris, computed π to 10,000 places, using an IBM 704, in 100 minutes.

1959. Genuys computed π to 16,167 decimal places, using an IBM 704, in 4.3 hours.

1961. Wrench and Daniel Shanks, of Washington, D.C., computed π to 100,265 decimal places, using an IBM 7090, in 8.7 hours.

1966. On February 22, M. Jean Guilloud and his co-workers at the Commissariat à l'Énergie Atomique in Paris attained an approximation to π extending to 250,000 decimal places on a STRETCH computer.

1967. Exactly one year later, the above workers found π to 500,000 places on a CDC 6600.

43° *The irrationality and the transcendentality of* π. It was in 1767 that J. H. Lambert showed that π is irrational, that is, that it is not of the form a/b, where a and b are integers, and in 1794 A. M. Legendre showed that π^2 is also irrational. In 1882, C. F. L. Lindemann proved that π is transcendental, that is, it is not a root of any polynomial equation of the form

$$a_0x^n + a_1x^{n-1} + \cdots + a_{n-1}x + a_n = 0,$$

where $a_0, a_1, \ldots, a_{n-1}, a_n$ are integers.

44° *The normalcy of* π. There is more to the calculation of π to a large number of decimal places than just the challenge involved. One reason for doing it is to secure statistical information concerning the "normalcy" of π. A real number is said to be *simply normal* if in its decimal expansion all digits occur with equal frequency, and it is said to be *normal* if all blocks of digits of the same length occur with equal frequency. It is not known if π (or even $\sqrt{2}$, for that matter) is normal or even simply normal. The fantastic calculations of π, starting with that on the ENIAC in 1949, were performed to secure statistical information on the matter. From counts on these extensive expansions of π, it would appear that the number is perhaps normal. The erroneous 707-place calculation of π made by William Shanks in 1873 seemed to indicate that π was not even simply normal.

The matter of the normalcy or nonnormalcy of π will never, of course, be resolved by electronic computers. We have here an example of a theoretical problem which requires profound mathematical talent and cannot be solved by computations alone. The existence of such problems ought to furnish at least a partial antidote to the disease of *computeritis*, which seems so rampant today. There is a developing feeling, not only among the general public, but also among young students of mathematics, that from now on any mathematical problem will be resolved by a sufficiently sophisticated electronic machine. These machines are merely extraordinarily fast and efficient calculators, and are invaluable only in those problems of mathematics where extensive computations can be utilized.

The elaborate calculations of π have another use in addition to furnishing statistical evidence concerning the normalcy or nonnormalcy of π. Every new automatic computing machine, before it can be adopted for day-to-day use, must be tested for proper functioning, and coders and programmers must be trained to work with the new machine. Checking into an already found extensive computation of π is frequently chosen as an excellent way of carrying out this required testing and training.

45° *Brouwer's question.* The Dutch mathematician L. E. J. Brouwer (1882–1966), seeking, for logical and philosophical purposes, a mathematical question so difficult that its answer in the following ten or twenty years would be very unlikely, finally hit upon: "In the decimal expression for π, is there a place where a thousand consecutive digits are all zero?" The answer to this question is still not known. It follows that the assertion that such a place in the decimal expression of π does exist is an example of a proposition that, to a member of the intuitionist school of philosophy of mathematics (Brouwer was a leader of this school), is neither true nor false. For, according to the tenets of that school, a proposition can be said to be true only when a proof of it has been constructed in a finite number of steps, and it can be said to be false only when a proof of this situation has been constructed in a finite number of steps. Until one or the other of these proofs is constructed, the proposition is neither true nor false, and the law of excluded middle is inapplicable. If, however, one asserts that a thousand consecutive digits are all zero somewhere in the first quintillion digits of the decimal expression for π, we have a proposition that is true or false, for the truth or falseness, though not known, can surely be established in a finite number of steps. Thus, for the intuitionists, the law of excluded middle does not hold universally; it holds for finite situations but should not be employed when dealing with infinite situations.

If π is normal, as is suspected, then a block of 1000 zeros will occur in its decimal representation not only once, but infinitely often and with an average frequency of 1 in 10^{1000}. It follows that much more is required to prove that π is normal than just to answer Brouwer's question, and this latter appears in itself to be a considerable task.

46° π *by legislation.* [The following is adapted, with permission, from the article "What's new about π?" by Phillip S. Jones, that appeared in *The Mathematics Teacher*, March, 1950, pp. 120–122.]

An often quoted but rarely documented tale about π is that of the attempt to determine its value by legislation. House Bill No. 246, Indiana State Legislature, 1897, was written by Edwin J. Goodwin, M.D., of Solitude, Posey County. It begins as follows:

> A bill for an act introducing a new mathematical truth and offered as a contribution to education to be used only by the State of Indiana free of cost by paying any royalities whatever on the same—
>
> Section 1. Be it enacted by the General Assembly of the State of

> Indiana: it has been found that a circular area is to the square on a line equal to the quadrant of the circumfcrcncc, as the area of an equilateral rectangle is to the square on one side. . . ."

The bill was referred first to the House Committee on Canals and then to the Committee on Education, which recommended its passage. It was passed and sent to the Senate, where it was referred to the Committee on Temperance, which recommended its passage. In the meantime the bill had become known and ridiculed in various newspapers. This resulted in the Senate's finally postponing indefinitely its further consideration, in spite of the backing of the State Superintendent of Public Instruction, who was anxious to assure his state textbooks of the free use of this copyrighted discovery. The detailed account of the bill together with contemporary newspaper comments makes interesting reading.

[We can incontestably establish the aberrant condition of the author of the bill by quoting the final section of the bill:

> Section 3. In further proof of the value of the author's proposed contribution to education, and offered as a gift to the State of Indiana, is the fact of his solutions of the trisection of an angle, duplication of the cube and quadrature of the circle having been already accepted as contributions to science by the *American Mathematical Monthly*, the leading exponent of mathematical thought in this country.
>
> And be it remembered that these noted problems had been long since given up by scientific bodies as unsolvable mysteries and above man's ability to comprehend.]

47° *Morbus cyclometricus.* There is a vast literature supplied by sufferers of *morbus cyclometricus*, the circle-squaring disease. The contributions, often amusing and at times almost unbelievable, would require a publication all to themselves. As a couple of samples, in addition to the one given in the preceding item, consider the following.

In 1892, a writer announced in the *New York Tribune* the rediscovery of a long lost secret that leads to 3.2 as the exact value of π. The lively discussion following this announcement won many advocates for the new value.

Again, since its publication in 1934, a great many college and public libraries throughout the United States have received, from the obliging author, complimentary copies of a thick book devoted to the demonstration that $\pi = 3\frac{13}{81}$.

There have been sufferers of *morbus cyclometricus* in rather high and responsible positions; in this country one sufferer was a college president, another was a member of the Lower House of the State of Washington, and still another served in the United States Senate.

GEMATRIA

SINCE many of the ancient numeral systems were alphabetical systems, it was natural to substitute the number values for the letters in a name. This led to a mystic psuedo-science known as *gematria*, or *arithmology*, which was very popular among the ancient Hebrews and others, and was revived during the Middle Ages. Part of later gematria was the art of beasting—that is, the occupation of cunningly pinning onto a disliked individual the hateful number 666 of the "beast" mentioned in the *Book of Revelations*. Item 130° of *In Mathematical Circles* describes Michael Stifel's arithmological effort to beast Pope Leo X, and Father Bongus's similar effort to beast Martin Luther. The Reformation led to a flood of such attempts against the Popes in Rome.

48° *Latin beasting*. The reader may care to check the following beastings, by considering only the letters representing Roman numerals, and taking U as V.

(1) LUDOVICUS (presumably Louis XIV)
(2) SILVESTER SECUNDUS (Gerbert, who reigned as Pope Sylvester II)
(3) PAULO V. VICE-DEO
(4) VICARIUS FILII DEI
(5) DOCTOR ET REX LATINUS
(6) VICARIUS GENERALIS DEI IN TERRIS
(7) DUX CLERI

49° *Greek beasting*. In Greek beasting one writes the name or title of the disliked person in Greek letters, and then employs the Greek alphabetical numeral system:

1	α	alpha	10	ι	iota	100	ρ	rho
2	β	beta	20	κ	kappa	200	σ	sigma
3	γ	gamma	30	λ	lambda	300	τ	tau
4	δ	delta	40	μ	mu	400	υ	upsilon
5	ε	epsilon	50	ν	nu	500	φ	phi
6	obsolete	digamma	60	ξ	xi	600	χ	chi
7	ζ	zeta	70	ο	omicron	700	ψ	psi
8	η	eta	80	π	pi	800	ω	omega
9	θ	theta	90	obsolete	koppa	900	obsolete	sampi

It was in this way that someone once beasted GLADSTONE; appropriate replacements give

γλαδστουη,

which achieves the beasting.

50° *Turning the tables.* The following two amusing beastings are found in De Morgan's *A Budget of Paradoxes.*

(1) "A Mr. James Dunlop was popping at the Papists with a 666-rifled gun, when Dr. Chalmers quietly said, 'Why, Dunlop, you bear it yourself,' and handed him a paper on which the numerals in IACOBVS DVNLOPVS were added up."

(2) "Mr. Davis Thom found a young gentleman of the name of St. Claire busy at the Beast number: he forthwith added the letters in στ χλαιρε and found 666."

51° *The greatest political figure.* Using gematria one can "prove" by simple English key that of the three men, Roosevelt, Churchill, and Stalin, Roosevelt was the greatest political figure.

52° *Amen.* The word "amen" when written in Greek is αμην. This is why, in certain Christian manuscripts, the number 99 appears at the end of a prayer.

53° *The Nile identified as the year.* Heliodorus (fourth century) claimed the Nile to be nothing else than the year, since the numbers expressed by the letters ΝΕΙΛΟΣ (Nile) are, in the Greek numeral system, Ν = 50, Ε = 5, Ι = 10, Λ = 30, Ο = 70, Σ = 200, and these add up to 365, the number of days in the year.

COUNTING BOARDS

MANY of the computing patterns used today in elementary arithmetic, such as those for performing long multiplications and divisions, were developed as late as the fifteenth century. Two reasons are usually advanced to account for this tardy development, namely the mental difficulties and the physical difficulties encountered in such work.

The first of these, the mental difficulties, must be somewhat discounted. The impression that the ancient numeral systems are not amenable to even the simplest calculations is largely based on a lack of familiarity with these systems. Still remaining, however, is the painful memorization of basic addition and multiplication tables, the learning of which occupies so much of our time in the elementary schools.

The physical difficulties encountered, on the other hand, were quite real. Without a plentiful and convenient supply of some suitable writing medium, any very extended development of arithmetic processes was bound to be hampered. It must be remembered that our common machine-made pulp paper is little more than a hundred years old. The older rag paper was made by hand, was consequently expensive and scarce, and, even at that, was not introduced into Europe until the twelfth century.

An early paperlike writing material called *papyrus*, tediously made from the papu reed, was invented by the ancient Egyptians, and by 650 B.C. had been introduced into Greece. But papyrus was too valuable to be used in any great quantity as mere scratch paper. Another early writing medium was parchment, made from the skins of animals, usually sheep and lambs. Naturally this was hard to get. Even more valuable was vellum, a parchment made from the skins of calves. Small but relatively inconvenient boards bearing a thin coat of wax, along with a stylus, formed a writing medium for the Romans of about two thousand years ago, and similar boards dusted with a flour were used in early India. Before and during the Roman Empire, sand trays were frequently used for simple counting and for the drawing of crude geometrical figures. Of course, stone and clay were used very early for making written records. Somewhat more convenient were the slates and slate pencils used in our schools within the past hundred years.

The way around these mental and physical difficulties was the invention of the *abacus* (Greek *abax*, sand tray) or *counting board*, which can be called the earliest mechanical computing device used by man. It appeared in many forms in parts of the ancient and medieval world, and was so widely employed up until almost the present that it made a deep impression in our culture. Some form or other of the abacus is still commonly used in various parts of the Orient. The basic idea of an abacus or counting board is to represent numbers by counters properly placed on a set of parallel lines, or by pebbles in a set of parallel grooves, or by beads on a set of parallel wires. An abacus is particularly adaptable for the representation of numbers of a positional numeral system.

54° "*The Hands.*" When the American military forces occupied Japan in 1945, the soldiers witnessed Japanese merchants and school children performing arithmetical calculations on the Japanese abacus, or soroban. The soldiers tended to ridicule the instrument for its supposed primitiveness. Feeling they should demonstrate the superiority of their modern methods, they organized a calculating contest in Tokyo, which was witnessed by some 3000 spectators. The outcome was surprising to the American soldiers.

A 22-year-old Japanese Communications Ministry clerk, Kiyoshi Matsuzaki, with seven years' special training with an abacus, was matched against 22-year-old Pvt. Thomas Ian Wood of Deering, Missouri, an Army finance clerk who had had four years' experience with modern electric desk calculators. Matsuzaki used an ordinary Japanese soroban, with a pre-war price of about 25 cents; Wood used an electric machine valued at $700.

Matsuzaki fluttered his hand over his instrument with such blurring dexterity that the Americans promptly nicknamed him "The Hands." In the contest, "The Hands" won all six heats of the addition event, finishing one of them a minute ahead of Wood. He also won in subtraction. Wood scored in multiplication, but "The Hands" won in division, and also in the final composite problem involving all four of the arithmetical operations. Also, "The Hands" made fewer mistakes.

55° *The return of the abacus to western Europe.* The French mathematician Jean-Victor Poncelet, in the rank of a lieutenant, accompanied Napoleon on his fateful 1812 invasion of Russia. During the French retreat, Poncelet was captured and taken to Saratov on the Volga. Living there for two years among simple people, Poncelet became impressed with the excellence of the Russian abacus as a device for teaching children. Upon his eventual release and return to France, he introduced the instrument into all the schools in the city of Metz, from where it spread all over France.

The abacus is recommended for use in our elementary schools of today because of its high pedagogical value in teaching positional notation and enumeration in different bases.

56° *The number tree.* The lines of a counting board can be drawn vertically or horizontally. In fact, no lines at all need to be drawn; their would-be positions can be indicated by a horizontal or a vertical row of counters instead. Thus, if the custom is to work with horizontal lines, these lines may be replaced by a vertical row of counters which are set down at the start and not moved again; these

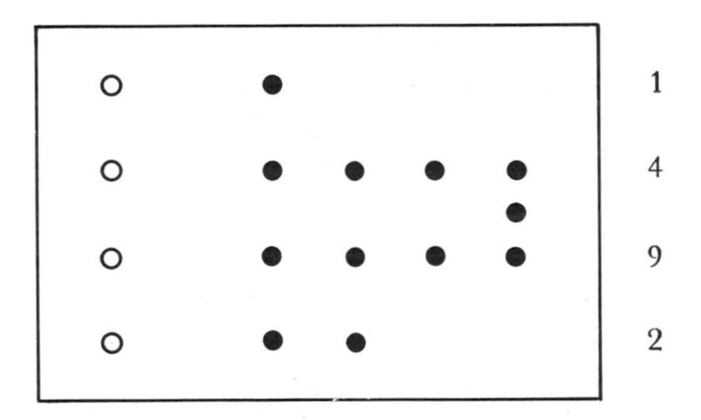

FIGURE 10

counters are to mark the positions where lines would have been. Such an initial set of fixed counters was called "the number tree," and was described in early French arithmetic books dating as far back as the fifteenth century. Figure 10 shows the number 1492 cast on a counting board on which the horizontal lines have been replaced by the number tree (the set of clear circles).

57° *The abacus as part of our culture.* In older days, a merchant had an abacus, or counting board, on a desk in his shop. Today one frequently refers to the table where the clerk of a store figures the customer's bill as "the counter."

A Roman abacus consisted of a board or a metal plate having a number of parallel grooves for the counting lines. The counters were small pebbles that were shifted along the grooves. The Latin word for pebble is *calculus*, from which our word "calculate" was derived.

Our present addition and subtraction patterns, along with the concepts of "carrying over" in addition and of "borrowing" in subtraction, in all likelihood originated in the processes for carrying out these operations on an abacus.

58° *Origin of the verb "to compute."* Karl Menninger, in his excellent *Number Words and Number Symbols, a Cultural History of Numbers* (The M.I.T. Press, 1969) has very neatly summarized the origin, in various languages, of the verb "to compute." The tracing of the root meanings of this verb leads to the oldest computing instruments—the fingers, the tally stick, and the counting board.

Language	Word	Original Meaning	Derived from
Greek	pempazein pséphizein	"to make fives" "to move pebbles"	finger counting counting board
Latin	computare calculos ponere	"to cut" "to place stones"	notched tally stick counting board
Medieval Latin	calculare	"to move pebbles"	counting board
French	jeter compter calculer	"to throw" < Lat. computare	counting board
English	to cast to count	"to throw" < Lat. computare	counting board

59° *A recurring simile.* In the second century B.C., the historian Polybius wrote:

> The courtiers who surround kings are exactly like counters on the lines of a counting board, for, depending on the will of the reckoner, they may be valued either at no more than a mere chalkos, or else at a whole talent!

It is to be borne in mind that the chalkos and the talent are the lowest and the highest values on the counting board.

The Polybius simile appears again and again. Here is an instance by Martin Luther, in the early sixteenth century:

> To the counting master all counters are equal, and their worth depends upon where he places them. Just so are men equal before God, but they are unequal according to the stations in which God has placed them.

Here it is again, by another (now unknown) sixteenth century writer:

> Life in this ephemeral world and all men in it are like a counter, which is worth so much (and no more) according to the line on which it lies and designates a sum. Now it lies on the top line. Why waste words? Before it can look around, the reckoner takes the counter away again, and again it is no more than just another counter, a mere piece of brass.

The mataphor appears again in the following lines of some mid-eighteenth century writer. The original verse was in French and has been attributed to several persons, including Frederick the Great.

> Courtiers are but counters;
> Their value depends on their place:
> In favor, they're worth millions,
> And nothing in disgrace!

60° *Das Hundert ins Tausend werfen.* The phrase "das Hundert ins Tausend werfen" ("throws the hundred into the thousand") is an expression still heard in Germany today. It is used to indicate confusion and lack of order, and is clearly inherited from the days of the counting board; anyone who places a counter, not on the 100-line of the board where it belongs, but on the 1000-line, creates confusion. On one occasion Martin Luther thundered: "The Devil is in a rage and he throws the hundred into the thousand, and thereby creates so much confusion that no one knows what to think."

61° *Our final accounting.* A Swabian saying goes: "We all shall have to make an account of ourselves on the red-hot counting board in God's chancery."

62° *A deception.* In imitation of coins, the custom arose (first in France) of making the counters used on a counting board out of metal and then stamping designs and inscriptions on them. Of course these counters were not real money, but they could be mistaken for money and were sometimes passed off as money—as metal bus tokens are sometimes passed off today. It was this deception that lay behind Martin Luther's remark: ". . . as if I were a child or a fool, who could be palmed off with counters instead of gulden." Some counters bore warnings of their falseness as coins.

63° *Political pamphlets of the day.* Appropriately inscribed counters frequently glorified some ruler or conqueror, or warned against some oppressor, and thus served as political propaganda. Counters were often struck as commemorative medallions.

TALLY STICKS

THE earliest way man kept a count was by some simple tally method, employing the principle of one-to-one correspondence. Such counts, if small, could be made on one's fingers. Larger counts could be maintained by making collections of pebbles or stones, by making scratches in the dirt or on a stone, by tying knots in a string, or by cutting notches in a piece of wood. This last method in time evolved into a widespread method of keeping records of debts and payments, and the notched pieces of wood came to be known as *tally sticks*.

64° *Some primitive tallies.* A primitive American Indian might keep count of the number of foes he has vanquished by collecting a scalp from each slain enemy. A primitive African hunter might keep count of the number of boars he has killed by collecting the boars' tusks. The young unmarried girls of the warlike tribe of Masai herdsmen who live on the slopes of Mt. Kilimanjaro keep an account of their ages by adding a brass ring around their necks each year. Children today often mark off the days to Christmas on the calendar.

The bartender of earlier days kept a tally of a customer's drinks by marks on a slate. At one time in Spain, for each drink the bartender would toss a small pebble (*chinas*) into the hood of the customer's cloak, whence the expression *echai chinas* ("to toss pebbles") still means "to chalk something up to someone's account."

The fourth-century Greek sculptor Lysippos claimed that he set aside one gold denarius for each of his pieces of work. When he died, he left his heir 1500 denarii. From this Pliny deduced that Lysippos produced a total of 1500 pieces of sculpture during his lifetime.

The German poet Adelbert Stifter has told, in one of his letters, how he kept count of the three weeks that had to pass before he could see his betrothed again. He secured 21 apples and ate one each day of the three-week separation.

65° *Tally sticks.* At one time it was fairly universal to keep records of debts and payments by means of properly notched sticks of wood, called tally sticks. Often a tally stick would be split lengthwise across the notches, the debtor receiving one piece and the creditor retaining the other.

It was in the British Exchequer that tally sticks achieved their highest development, serving as official government records. A number of years ago, when repairs were being made in Westminster Abbey, several hundred exchequer tallies of the British Royal Treasury were uncovered. These tallies dated from the thirteenth century and recorded amounts of tax owed and paid. Such tallies continued in use until 1826, the method of notching the sticks remaining the same until that year. The official rules for notching the sticks have come down to us in the *Dialogus de Scaccario*, written in 1186 by the Royal Treasurer Richard, Bishop of London.

> The notch for £1000 is placed at the end and is as large as the hand is wide [see Figure 11(a)];
> for £100 the notch is as large as the thickness of a thumb, and to

distinguish it from that for £1000 it is not straight but curved [see Figure 11(b)];
for £20 it is as large as the thickness of the little finger [Figure 11(c)];
for £1 it has the breadth of a ripe barleycorn [Figure 11(d)];
for 1 shilling it is smaller but still large enough to be seen as a notch [Figure 11(e)]; whereas
for 1 penny only a cut is made, with no wood being removed [Figure 11(f)];
for a half of any of these units a notch or cut half the length is carved: one cut slantwise, one perpendicular to the edge (Figure 11(g)).

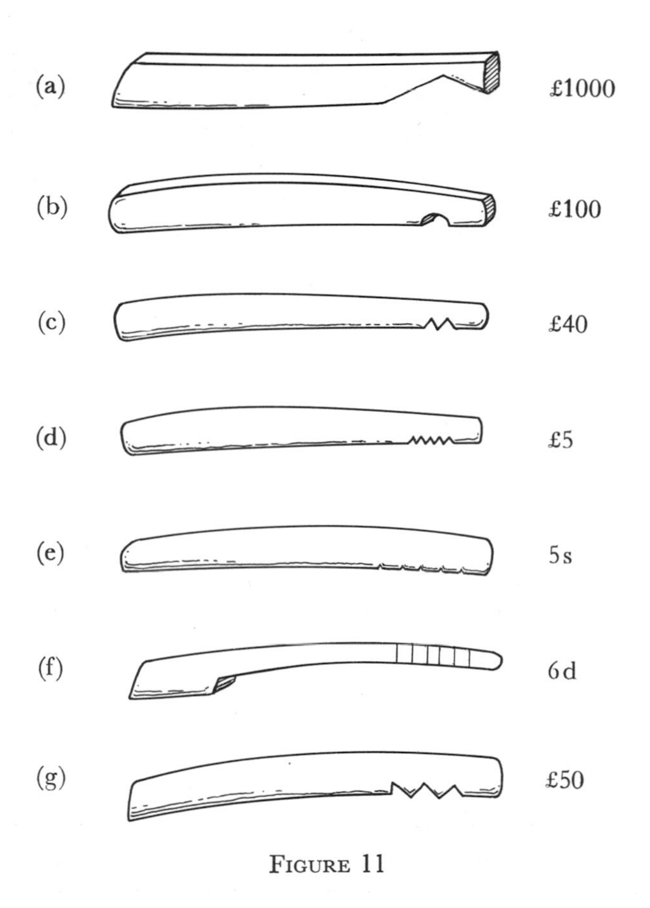

FIGURE 11

66° *Pictographic evidence of the tally stick and the counting board in Chinese culture.* Figure 12(a) shows a Chinese pictograph for the word "contract." It is symbolized by combining three characters. The two on top are characters for "tally stick" and "knife" (for notching the tally stick); the one below is a character for "large." Thus a "contract," or "agreement," in Chinese is literally "a large tally stick."

A Chinese character for "book" is shown in Figure 12(b). It is a bundle or cluster of tally sticks strung together, reminding one of just such a Russian tax book in the Kansallis Museum of Helsinki. The word "law" is then represented by a book of tally sticks resting on a table, as shown in Figure 12(c).

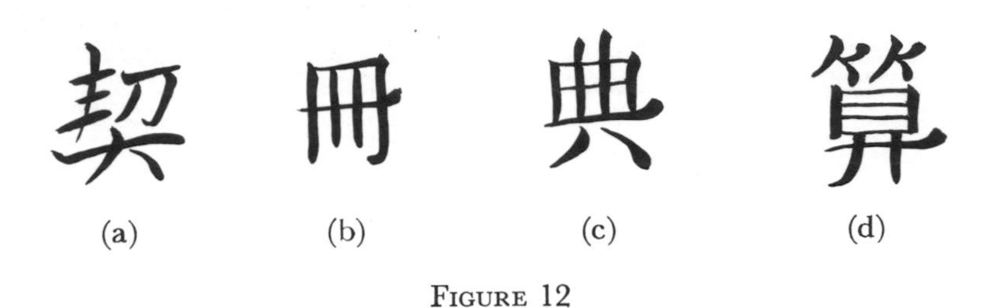

FIGURE 12

Figure 12(d) is a Chinese pictograph for the verb "to calculate." This character also stands for "suan-pan," the Chinese counting board, and is compounded from a character for two "hands" (at the bottom), holding a "reckoning board" (in the middle), made of "bamboo" (at the top).

67° *Tally sticks and the burning of Parliament.* Tally sticks remained in use in the British Exchequer for many centuries. In 1782 it was ordered that tallies no longer be employed, but existing ones remained valid until 1826. In 1834 the state's official collection of cancelled tallies was disposed of by burning in the furnaces beneath the Houses of Parliament. The furnaces became so overheated that the Parliament buildings caught fire and were destroyed.

Charles Dickens has added a wry note to the above accident. It had been urged by some that the cancelled tally sticks be given as firewood to the poor people living in the neighborhood of the Houses of Parliament. The government was unwilling to do this, and in its zeal to destroy the sticks it managed to burn down the Parliament buildings.

The government then went around, to those same poor people, to collect extra taxes for the rebuilding of the burned structures.

68° *A simple tally stick.* Karl Menninger, in his fascinating *Number Words and Number Symbols* (The M.I.T. Press, 1969), quotes the following anecdote about the Flemish painter Pieter Brueghel (d. 1569) as given by one of the artist's contemporaries:

> While he was in Antwerp, he lived with a young girl whom he would have married if she had not had the unfortunate habit of constantly telling lies. He made a pact with her, that for each lie she told he would cut a notch in a piece of wood; he took a good long stick, and when the stick was filled with notches, there could be no longer any question of marriage; this happened before very long.

69° *The English word "score."* There is evidence that the word *score* originally meant the scar or notch that a farmer would cut in a stick for every twenty cattle or sheep that he was counting. Later it came to represent the notch used to indicate £20 on the British exchequer tallies, and thence came such expressions as: What is the score? He has paid his score to society.

In Shakespeare's *King Henry VI*, Part Two, Act 4, Scene 2, we find:

> I thank you, good people. There shall be no money, all shall eat and drink on my score.

And, from Scene 7 of the same act, we have

> Our forefathers had no other books than the score and the tally.

In Act 5, Scene 7, of *Macbeth*, it is said of Macbeth at the conclusion of the tragedy:

> He parted well and paid his score.

70° *Etymology of some business terms.* Our commonly used business words "stock" and "check" originated with the old English custom of using double tally sticks. When a man lent money to the Bank of England, the amount was cut on a tally stick. A splint or foil was then split from the stick and retained by the bank, while the lender was given the heavier, or stock, piece of the stick. In this way the lender became a "stock holder" and owned "bank stock." When the lender desired the return of his money, he had to "check" his stock with the foil kept by the bank. The expression, "our accounts tally," also stems from the comparison of stock and foil.

71° *Multiple tally stick.* Not very long ago, Viennese snow-removal men used a three-part tally stick—that is, a tally stick from which two splints were split, one from each side of the central stock. The middle part was kept by the driver, one of the side pieces was given to the foreman at the loading platform, and the other side piece was given to the foreman at the unloading platform. The result was a record in triplicate, much like a present-day record with two carbon copies.

With double and triple tally sticks, cheating was impossible. For this reason such tally sticks had the force of legal documents.

The idea of the double tally stick was carried over to written documents as early as the tenth century. Two copies of a contract would be written, one below the other, on a piece of paper or parchment, and then fancy letters drawn in the space between the two texts. The two records were then separated from one another by a jagged or wavy cut, and each party to the contract took a piece. Such documents were called "divided papers" or "toothed papers." Today we achieve the same end by using carbon paper.

COMPUTERS

FIELDS in which numerical calculations are important, such as science, navigation, business, engineering, administration, and war, have made ever increasing demands that these computations be performed more quickly and accurately. These increasing demands were met successively by four remarkable inventions: the Hindu-Arabic notation, decimal fractions, logarithms, and modern computing machines.

Most of the early electronic computers were designed to solve military problems, but today they are being designed for science, business, industry, administration, government, and other purposes. From luxury tools they have become vital and necessary instruments in present-day development. It is now not uncommon for secondary schools to offer introductory courses in computer science and to have a tie-up with some large computer located at a nearby college or university. Babbage's dream (see the historical capsule immediately preceding Item 275° of *In Mathematical Circles*) has indeed come true!

Perhaps motivated by a certain feeling of affright at the incredible ability of the modern computers, there is a tendency to level unjustified and unearned gibes at the machines.

72° *An analogy.* Just as the bulldozer has greatly extended man's muscles, the modern high-speed electronic computer has extended man's brains. Anything that is done by a bulldozer could be done by a sufficiently large group of laborers working a sufficiently long period of time, but the task might be so prolonged and unpleasant without the bulldozer that it would never be undertaken. Similarly, anything that is done by a high-speed computer could be done by a sufficiently large group of arithmeticians working a sufficiently long period of time, but the task might be so arduous and so unpleasant without the computer that it would never be undertaken. Just as the bulldozer revolutionized certain features of our lives, so will the computer revolutionize others, and we will now undertake tasks that we would scarcely have dreamed of in the past.

73° *Speed.* The IBM 7090 electronic digital computing machine can, in one second, perform 229,000 additions or subtractions, or 39,500 multiplications, or 32,700 divisions.

74° *A cartoon by P. Barlow.* Two worried administrators stand before a large electronic computer, examining a piece of paper containing the machine's latest output. One administrator in exasperation says to the other: "Old Miss Pringle at her worst never made a mistake like this."

75° *A cartoon by J. Mirachi.* Again two administrators stand before a great electronic computer, reading the machine's latest analysis of an involved linear programming problem. One administrator says to the other: "Still the same answer. To increase the margin of profit, put more bread crumbs in the hamburger."

76° *Watch out.* There's a story going the rounds about a Dutch professor who fed this question into a very sophisticated computer: "I have the choice between two watches; one is broken and irrevocably stopped, the other loses one second per 24 hours. Which watch should I buy?" The computer's reply: "The one that is stopped, as it indicates the correct time twice every 24 hours; the other does so only once every 120 years."

77° *The supercomputer.* "It's premature to worry that the computer is replacing the human brain," said Margaret McNamara, wife of Defense Secretary Robert McNamara. "The brain," she reminded us, "is the most magnificent bit of miniaturization in the universe. Though it weighs only three pounds, it contains some ten billion nerve cells, each of which has some 25,000 possible interconnections with other nerve cells. To build an electric computer large enough to have that range of choice would require an area equal to the entire surface of the earth."

78° *And cheaper to produce.* It has been pointed out that not only is the human brain the greatest computer ever devised, but it is the only one produced by unskilled labor.

WEIGHTS AND MEASURES

UNITS of length, area, volume, weight, and currency play an important part in the practical applications of arithmetic, and there is much quaint lore connected with the various units of measure.

79° *Concerning lengths and areas.* Many of the early units of length arose naturally from different parts of the human body. For example, in ancient Egypt and Sumer, a convenient small unit of length was the *cubit*—the distance from a man's elbow to the tip of his extended middle finger. Of course, this unit varied with the individual. The same was true of such units as the *foot*, the *hand* or *palm* (the width of the human hand), the *span* (the distance from the end of the thumb to the end of the little finger of an outstretched hand), the *finger* (the width of the index finger), and the *yard* (the distance from the tip of the nose to the end of an outstretched arm).

To obtain a uniform unit, a *yard* in England was later decreed to be the distance from the tip of the nose to the end of the outstretched arm of King Henry I. The word "yard" comes from the Anglo-Saxon "yerde," meaning stick, and today we frequently speak of a yardstick. Cloth merchants used to measure cloth by stretching it, as it was unrolled from the bolt, from nose to outstretched hand. Because of the stretching, a legal yard became a cloth yard "and a handful." In 1439, it became a cloth yard "and a thumb." A *nail* is the sixteenth part of a yard, or $2\frac{1}{4}$ inches; it is now obsolete but was used in measuring cloth. A *hand*, now used almost solely for measuring the heights of horses, has become standardized as four inches. A *span* is half a cubit.

An *ell*, which is a measure of length little used today, varies considerably from country to country, and has "arm" or "forearm" for its original meaning. An ell would thus seem to be a cubit. The English ell became standardized at about 45 inches—a remarkably long forearm indeed.

The Romans used a duodecimal system and accordingly divided a foot into twelve equal parts, called "unciae" (twelfths), from which our word *inch* has derived. In 1324, King Edward II of England standardized the inch as the length of "three barleycorns, round and dry, and laid lengthwise."

In printing, a *line* is one twelfth of an inch, and an *em* is one sixth of an inch. Originally an em was the width of the upper case letter M in the type faces. An *en* is half an em, and linotype operators measure a line of type in ens.

A surveyor's *chain* became standardized at twenty-two yards, and was made up of a hundred *links*. A *pole* or *rod* originally was a measuring stick, equal to a double pace, and was $5\frac{1}{2}$ yards, or one fourth of a chain, long. In 1514 a rod was decreed to be the sum of the lengths of the left feet of the first sixteen men coming out of church at the end of the service. An area of one square pole ($30\frac{1}{4}$ square yards) is called a *perch*. An *acre* was supposed to be as much land as a team of oxen could plow in a day (or perhaps in a morning); it became standardized at ten square chains, or 4840 square yards. A fourth of an acre ($2\frac{1}{2}$ square chains) is called a *rood*.

A *furlong* originally meant the length of a furrow in plowing; it is now one eighth of a mile, or 220 yards. A *mile*, in turn, comes from the Latin "mille passuum," meaning a thousand double paces. The Roman army marching pace, like the present-day Boy Scout's pace, was said to be thirty inches, whence a Roman mile would be five thousand feet. The mile has considerable variation from country to country today. A *league* is another distance unit that varies much about the world today; in England a league at one time was taken as twelve furlongs, but is now considered as roughly three miles.

A *fathom* is two yards, and originally designated the spread, or embrace, achieved by extending one's arms horizontally sideways from the body. Fathoms came to be used mostly in measuring sea depths. A *cable's length* is strictly a tenth of a sea, or nautical, mile, and a *sea mile* was about two hundred yards or a hundred fathoms. Today a cable's length varies from 100 to 140 fathoms. The British navy cable is $12\frac{1}{2}$ fathoms (called a *short cable*); a length of 15 fathoms is called a *long cable*.

80° *Measure for measure.*

Teacher. King Henry the First decreed that the yard should be the distance from the tip of his nose to the end of his outstretched arm.

Student. Is that why they called him the ruler of England?

ALAN WAYNE

81° *Knots that are not knots.* A nautical *knot* is a speed and not a distance; it is used only for measuring speeds of ships. A knot is one nautical, or sea, mile per hour. A ship's log line had knots tied into it at equal intervals of $\frac{1}{120}$ of a sea mile. After the log line was thrown over the aft part of the ship, a sailor would count the number of knots on the log line which passed over the end of the ship in thirty seconds; this gave the ship's speed in knots or nautical miles per hour.

82° *Concerning volumes.* A *barrel* is a variable measure of capacity. Thus a barrel of wine contains $31\frac{1}{2}$ gallons, and a barrel of ale contains 36 gallons; a dry barrel may contain almost any quantity. The word *barrel* comes from the French "baril," which, in turn, comes from the Latin "barre," meaning bar or stave, for barrels were made of wooden staves. A *hogshead* is a large measure of wine or beer, not always the same, but often taken as 63 gallons. A *kilderkin* is an old beer measure equal to half a barrel, or 18 gallons. A *firkin* is a fourth of a barrel; if it is a firkin of ale or beer, it is nine gallons.

A *bushel* is a dry measure, made up of four *pecks* or 32 (dry) *quarts*. The British bushel, however, is larger than the United States bushel. *Sack*, *coomb*, *chaldron*, *wey*, and *last* are associated measures no longer in use, but formerly defined as follows:

1 sack = 3 bushels
1 coomb = 4 bushels
1 chaldron = 36 bushels
1 wey = 40 bushels
1 last = 80 bushels

The size of a sack sometimes varied. The chaldron was used almost exclusively for measuring coal, coke, and lime. A chaldron of coal weighed about $1\frac{2}{7}$ tons.

A *gallon* is a liquid measure, but, as with the bushel, a British and a United States gallon are different. An *imperial gallon* of England is equal to 277.42 cubic inches; a United States gallon is equal to 231 cubic inches. In either case, a gallon contains four *quarts*, a quart contains two *pints*, and a pint contains four *gills* or *quaterns*. The word pint came from the Latin "picta" (painted, or marked), through the Spanish "pinta" (a spot or a mark), to the French "pinte." Thus the word came to mean a mark painted on the side of a vessel to show the level of one pint. Old tables record a *glass* as one sixth of a pint, but later tables record it as one eighth of a pint. In the latter case, a *noggin* was considered to be two glasses—that is, a noggin is the same as a gill.

The standard United States gallon formerly was the British "old wine gallon," which was defined in two ways: as the content of a cylinder seven inches in diameter and six inches deep, or as a volume of 231 cubic inches. It is interesting to note that these two definitions are equivalent if and only if π is taken to be exactly 22/7.

83° *Concerning weights.* The *pound* is a unit of weight, varying with periods and with countries. In the United States and in the British Commonwealth, the pound has become either of two legally fixed units, the *pound avoirdupois* (of 7000 *grains* and divided into 16 *ounces*) used for weighing ordinary commodities, and the *pound troy* (of 5760 grains and divided into 12 *ounces*) used for weighing gold, silver, precious stones, and medicines.

The word "ounce" has the same etymological origin as the word "inch." The Romans divided their pound, as they did their foot, into twelve equal parts, called "unciae" (twelfths), from which both our words "inch" and "ounce" have derived.

A *stone*, which was once variable, became standardized in England as 14 pounds. As a larger unit of weight there is the *hundredweight*, which is equal to 100 pounds avoirdupois in the United States and to 112 pounds in England. One fourth of a hundredweight is called a *quarter*. In the United States a *ton* is 2000 pounds, and is sometimes called a *short ton*. A *long ton*, used in England, Australia, and American Customs houses, is 2240 pounds.

The *tare* is the weight of the wrapping or receptical containing goods; often a deduction of this from the gross weight is allowed. *Tare* is also the weight of a vehicle without cargo or passengers, and in chemistry it is a counterweight used to balance the weight of a container.

In England, a bundle of old hay containing 56 pounds, or new hay containing 60 pounds, or straw containing 36 pounds, is called a *truss*. Thirty-six trusses constitute one *load*.

A *pennyweight* is a troy weight of 24 grains, or $\frac{1}{20}$ of an ounce. It originally was the weight of the Old English silver penny, valued at 3 denarii.

84° *Tonnage*. The *tonnage* of a ship is not a measure of the ship's weight, but is 100 times its capacity in cubic feet below decks. Thus the Queen Mary, of 80,000 tonnage, had a capacity of 8,000,000 cubic feet below decks.

85° *Concerning former British currency*.* The English *penny* (plural *pence*) was related to the German "pfennig," and the old Anglo-Saxon penny corresponded to the Roman "denarius."

*On February 15, 1971, the United Kingdom changed from the currency system mentioned here and in Items 99 and 100 to one based on the decimal system.

A *farthing* was a bronze coin worth one fourth of a penny. The name "farthing" derives from Old English "fower," meaning four; the Middle English term was "ferthing."

A *shilling* was 12 pence, or $\frac{1}{20}$ of a pound sterling. The word "shilling" is of uncertain origin, but there is the similar "schilling" of Germany and "skillig" of Denmark.

Many European countries have had a coin called a *florin*. The first florins were made in Florence, Italy, and the coin secured its name from the lily that was imprinted on it, the lily being the symbol of the town of Florence. At the time of Edward III (1327–1377), the English florin was worth 6 shillings. The later English florin, which was first minted in 1849, was worth 2 shillings.

A *crown* was a silver coin worth 5 shillings, and it attained its name because it was stamped with a picture of a royal crown. A half-crown, of course, was worth 2 shillings and 6 pence.

A *guinea* was an Old English coin which at first had an approximate value of 20 shillings, but after 1717 had the fixed value of 21 shillings—an auctioneer was supposed to retain the odd shilling. The guinea received its name from the fact that in the beginning, in 1663, it was made from gold obtained from the Guinea Coast in Africa. The minting of guineas was abandoned about 1813.

The British *pound* was worth 20 shillings, or 240 pence. It was originally equivalent to a pound of silver.

A *sovereign* was an English gold coin worth 2 pounds and 18 shillings, and was intended only for use abroad. It received its name from the fact that it was stamped with the head of a British sovereign.

86° *Early programs for obtaining a standard unit of length.* Suggestions for obtaining a standard unit of length were proposed prior to the time the French National Assembly decided to commence work on the metric system.

In 1670, the French mathematician and vicar of the church of St. Paul at Lyons, Abbé Gabriel Mouton, suggested one minute of the earth's circumference as a length unit, and he divided and subdivided this unit decimally, assigning appropriate Latin terminology to the various divisions and multiples. About the same time, Sir Christopher Wren, in England, proposed taking the length of a pendulum beating half-seconds as the unit of length; this would have approximated one half of the length commonly assigned to the ancient cubit. In 1671 the French astronomer Jean Picard, and in 1673 the Dutch physicist Christiaan Huygens, advocated the length of a seconds pendulum at sea level on the 45th degree of latitude; this would have been only about 6 millimeters shorter than the present-day meter. In 1747, La Condamine suggested the seconds pendulum at the equator. In 1775, Messier very carefully determined the length of a seconds pendulum at 45° of latitude, and unsuccessfully endeavored to have this adopted as the standard linear unit.

It was in 1789 that the French Académie des Sciences appointed a committee to work out a plan for securing a new system of measures. In the following year Sir John Miller proposed, in the House of Commons, a uniform system of measurement for Great Britain. About the same time Thomas Jefferson proposed a uniform system for the United States, suggesting the length of a seconds pendulum at 38° of latitude, this being the mean latitude for the United States of his day.

In 1790, the French National Assembly, motivated by the widespread agitation for a new system of measures, decided to proceed immediately with a project of uniformity. The seconds pendulum was abandoned, and an arc of 1/10,000,000 of a meridional quadrant was selected as the fundamental unit of length. Today the standard meter is defined in terms of some selected wave length of light.

87° *A calamitous mistake.* Pierre François André Méchain must be ranked as one of the most hapless of mathematicians. He was born at Laon, August 16, 1744, and started his professional career as a private tutor of mathematics. He studied astronomy in his free time and finally attracted the interest of the noted French astronomer Joseph Lalande, who helped secure him a government post involving astronomical observations and the survey of the French coast. In 1782 he was elected a member of the French Academy, and in 1785 he became editor of *Connaissances des Temps*, in which some of his most valuable scientific papers appeared. In 1791 he was commissioned, with Cassini and Legendre, to find the difference of latitude between Dunkirk and Barcelona, as part of the program of determining the standard meter (1/10,000,000 of a meridional quadrant) as the fundamental unit of length. Working with Delambre, Méchain was assigned the task of measuring the part of the meridian between Rodez and Barcelona. The work was completed in 1799, when Méchain sent in his report to Paris. After he had submitted the report, he discovered that he had made an error of 3″ in calculating the latitude of Barcelona. In an effort to protect his scientific reputation, he concealed the error and sought to extend the meridian to the Balearic Islands, thus cutting out Barcelona altogether. While carrying out this plan he died of yellow fever at Castellon de la Plana, near Valencia, on September 20, 1804. The tragic error became known, with the result that instead of being considered a scientist of ability and repute, he became known as the man who made the chief mistake in the determination of the standard meter. In all fairness to Méchain, it should be said that the fault was hardly his, for obstacles were placed in his way that made accurate work almost impossible.

88° *A piece of neutral land.* As a site for the International Bureau of Weights and Measures, the French government assigned a piece of ground at Sèvres, in the Park of Saint Cloud, and declared it to be neutral territory. The expenses of the Bureau are borne by contributions from the contracting governments. The routine work of the Bureau is performed under the supervision of the International Committee of Weights and Measures, which, in turn, is subject to the control of a general conference to which all contracting governments periodically send delegates.

89° *Errors.* In 1821, a replica standard kilogram, made of platinum, was sent to the United States by Albert Gallatan, then Minister to France. This kilogram was authenticated, by a certificate from the famous physicist D. F. J. Arago, to differ from the official kilogram of the Archives by less than 1 milligram. However, when compared in 1879 with the British platinum kilogram, the United States standard was found to be 4.25 milligrams light. This discrepancy was later confirmed, in 1884, when the United States kilogram was taken to the International Bureau of Weights and Measures at Sèvres and compared with two auxiliary kilograms each known to have the same value as the kilogram of the Archives. This final determination yielded an error of −4.63 milligrams.

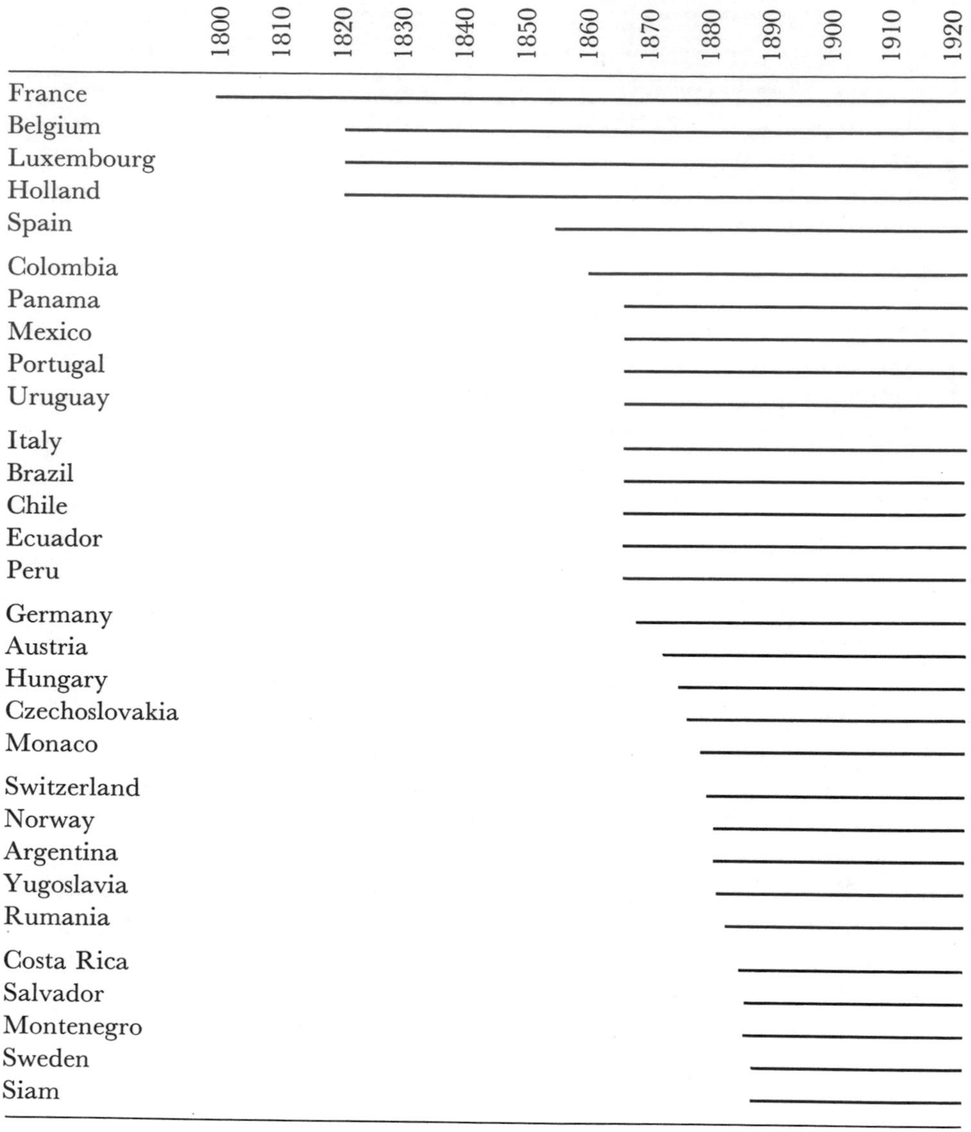

1800 1810 1820 1830 1840 1850 1860 1870 1880 1890 1900 1910 1920

France
Belgium
Luxembourg
Holland
Spain

Colombia
Panama
Mexico
Portugal
Uruguay

Italy
Brazil
Chile
Ecuador
Peru

Germany
Austria
Hungary
Czechoslovakia
Monaco

Switzerland
Norway
Argentina
Yugoslavia
Rumania

Costa Rica
Salvador
Montenegro
Sweden
Siam

FIGURE 13

Quadrant One [89°]

1800 1810 1820 1830 1840 1850 1860 1870 1880 1890 1900 1910 1920

Bulgaria
Egypt
Finland
Nicaragua
Bolivia

Guatemala
Tunisia
Honduras
Haiti
Paraguay

Cuba
Philippines
Belgian Congo
Denmark
Venezuela

Turkey
Latvia
Poland
Greece
Estonia

Lithuania
China
Russia
Japan
Morocco

British C'wealth
United States

Figure 13

Incidentally, the dollar worth of platinum has greatly increased since the offical platinum kilograms were made. Valued at about $300 at the time of construction, a kilogram of platinum today is worth several thousand dollars.

90° *Advance of the metric system.* On June 22, 1799, following the French Revolution, the Republic of France adopted the metric system of weights and measures. Of course, it took some time in France before the new system was widely used; the system was merely permissive until 1840, when it became legally mandatory. Year after year, following the French initiation in 1799, other countries of the world have adopted the metric system, until, by 1920, all countries of the world except the United States and the British Commonwealth have adopted it. The exceptional countries use the metric system in their science, but not for everyday common use.

The chart in Figure 13, adapted from a similar one in R. J. Gillings, *A Brief History of British Weights, Measures, Signs, Symbols, and Decimal Currency*, shows the advance of the adoption of the metric system.

QUADRANT TWO

From plus and minus
to Protestant curves

SYMBOLS AND TERMINOLOGY

APPROPRIATE symbolism and terminology play a cardinal role in mathematics, and there are many interesting, curious, or amusing stories connected with adopted or proposed notation and nomenclature.

91° *Whitehead on + and −.* In his peerless little book, *An Introduction to Mathematics* (1911), Alfred North Whitehead (1861–1947) comments on the + and − signs as follows: "There is an old epigram which assigns the empire of the seas to the English, of the land to the French, and of the clouds to the Germans. Surely it was from the clouds that the Germans fetched + and −; the ideas which these symbols have generated are much too important to the welfare of humanity to have come from the sea or from the land."

92° *Gauss on* $+1$, -1, *and* $\sqrt{-1}$. Commenting on the unfortunate use of the term "imaginary" in connection with complex numbers, Gauss, in his *Theoria residiorum biquadraticorum*, says: "That this subject has hitherto been considered from the wrong point of view and surrounded by a mysterious obscurity, is to be attributed largely to an ill-adapted terminology. If, for instance, $+1$, -1, and $\sqrt{-1}$ had been called *direct*, *inverse*, and *lateral* units, instead of *positive*, *negative*, and *imaginary* units, such an obscurity would have been out of the question."

93° *An odious notation.* Gauss, writing to his astronomer friend H. C. Schumacher, complained of the notation $\sin^2 \phi$ for $(\sin \phi)^2$: "$\text{Sin}^2 \phi$ is odious to me, even though Laplace made use of it. . . . Let us write $(\sin \phi)^2$, but not $\sin^2 \phi$, which by analogy should signify $\sin(\sin \phi)$." When John Herschel employed the notation $\sin^{-1} \phi$, $\tan^{-1} \phi$, etc., for the inverse trigonometric functions, he observed that the notation is at variance with $\sin^2 \phi$ for $(\sin \phi)^2$. He claimed that we ought to reserve $\sin^2 \phi$ for $\sin(\sin \phi)$, $\log^2 x$ for $\log(\log x)$, etc., in line with the custom of writing d^2x and $\Delta^2 x$ for ddx and $\Delta\Delta x$. He justified his notation for inverse functions since it was already customary to write $d^{-1}V$ for $\int V$, $d^{-2}V$ for $\int\int V$, etc.

94° *Figuratively speaking.* At James Madison High School in Brooklyn, Murray Navon was starting to teach a geometry class, and used the symbol: "∴".

"What is that?" asked a student.

"Therefore," explained Mr. Navon.

"You mean 'they're three,' don't you?" asked the student.

ALAN WAYNE

95° *The earliest symbol in mathematical logic.* [The following is adapted, with permission, from the article, by Howard Eves, of the same title that appeared in the Historically Speaking section of *The Mathematics Teacher*, January, 1959, p. 33.]

Although Gottfried Wilhelm Leibniz (1646–1716) is generally regarded as the founder of symbolic logic, some mathematical symbols of a logical, rather than of an operational, nature were employed before him. Thus Pierre Hérigone published, in Paris in 1634, a six-volume work on mathematics called *Cursus mathematicus*, in which he employed an extensive set of mathematical symbols of both the logical and the operational type. But his logical symbols were merely abbreviations of certain useful phrases. Thus he used "hyp." as an abbreviation for "from the hypothesis it follows," and "constr." for "from the construction one has," etc.

Perhaps the first genuine symbol of a logical nature, and one that has lasted into present times, is the familiar three dots, ∴, for "therefore." This symbol was introduced by Johann Heinrich Rahn (1622–1676), a Swiss mathematician who wrote in German, in his *Teutsche Algebra*, published in Zurich in 1659. In fact, in the *Teutsche Algebra* we find both ∴ and ∵ appearing for "therefore," the former symbol, however, predominating.

Rahn's treatise had only a minor influence in continental Europe. But in 1668 an English translation of the work, entitled *An Introduction to Algebra*, appeared in London, and this translation considerably influenced subsequent British writers. The translation was made by

Thomas Brancker (1636–1676) and was edited with alterations and additions by John Pell (1611–1685). As in the original treatise, both $\therefore$ and $\because$ appear as symbols for "therefore," but now $\because$ is predominant.

It is interesting to note that in Rahn's book, and in many later British texts of the eighteenth century, the two three-dot symbols for "therefore" were used principally in passing from a proportion to the equation obtained by equating the product of the extremes to the product of the means.

Today, in Great Britain and the United States, the symbol $\therefore$ is commonly used for "therefore," and the symbol $\because$ has come to mean "because." This differentiation in the meanings of the two symbols apparently did not occur until the beginning of the nineteenth century, and the symbol for "because" has not met with as wide an acceptance as that for "therefore." Both symbols are rarely found in continental European publications except in some on symbolic logic.

96° *A symbol for division.* The difference in influence of Rahn's original treatise, *Teutsche Algebra*, and its later English translation, may be noted by the fact that though Rahn also first introduced the symbol $\div$ for division, this symbol is so used today only in Great Britain and America. The symbol $\div$ is commonly used in continental Europe to indicate subtraction.

97° *Misplaced credit.* Another historical curiosity can be traced to the English translation of Rahn's *Teutsche Algebra.* Among the additions made by John Pell to the English translation was a laborious treatment, due jointly to John Wallis (1616–1703) and Lord Brouncker (ca. 1620–1684), of the diophantine equation $ax^2 + 1 = y^2$, where a is a nonsquare integer. Although Pell had no other connection with the equation than this, because of a misunderstanding on the part of the great Swiss mathematician Leonhard Euler (1707–1783) the equation has become known in number theory as *the Pell equation.*

98° *Pound sterling and pound weight.* School children sometimes express curiosity over the respective representations £ and lb. for the British pound sterling and the pound weight. Both symbols originate from the word "libra" for the ancient Roman pound.

Since a British pound sterling was originally equivalent to a pound of silver, it came to be denoted by the initial capital letter L of Libra. The symbol at first had only one horizontal stroke through it, but today it is sometimes printed with two strokes. In fact, in the early nineteenth century, the lower case "l," without any stroke, was frequently used. It is thought that the reason for the stroke lay in the need to distinguish the letter L for Libra from the Roman numeral L for 50. In France there is the "livre" and in Italy there is the "lira," also derived from the Latin "libra."

The sign "lb." for pound weight comes from the first and third letters of the Latin word "libra." At one time, this sign had a stroke through it, like in £, and it is still sometimes seen this way in England. In an early edition of the *Encyclopaedia Britannica*, the sign is printed with a stroke, as ℔.

99° *Penny and pennyweight.* The signs for the former British penny and pennyweight were d. and dwt. respectively. The d. came from the word "denarius," which was a more or less equivalent coin in Roman days.

100° *Shillings.* A *solidus* was a Roman gold coin introduced by Constantine. It was abbreviated by the letter "s," which in time became elongated into a sloping stroke (/) called a solidus. The solidus coin was introduced into medieval Europe as a piece of money evaluated at 12 denarii. The Anglo-Saxon coins corresponding to the solidus and the denarius were the shilling and the penny, a shilling being equal to 12 pence. Thus the solidus stroke came to represent shillings when one wished to separate shillings from pence, as in 3/6 for 3 shillings and 6 pence.

101° *Ounce.* The first letter of the abbreviation "oz." for ounce can be accounted for as the first letter of the word *ounce*. But whence the "z"? It seems that in early-California mining days, there was a Spanish coin called "onza de oro," which contained one ounce of gold. The "z" is from the Spanish word "onza."

102° *Percentage.* The sign % for percentage seems to have taken many years to develop. It went through a succession of stages: p. ceto, p.$^{o}_{c}$, p. 100, and in the seventeenth century it was written as p$\frac{o}{o}$. The symbol ‰ has been suggested to signify *per mille*.

103° *Backward people.* Sometimes we in America think that the British are a funny lot; they drive on the wrong side of the road and they hold their forks with the wrong hand. Also, they place their points for multiplication on the line, and their decimal points half-way up—just the reverse of what we do in the United States. For that matter, in France and other parts of Europe a comma is used in place of our decimal point; it is called the *virgule decimale*, or decimal comma.

104° *A brief history of the dollar mark.* [The following is adapted, with permission, from the article of the same title by Sharon Ann Murray that appeared in the Historically Speaking section of *The Mathematics Teacher*, October, 1959, pp. 478–479.]

Contrary to widespread belief, our dollar mark did not originate in the combination of the letters U and S. The symbol actually has a Spanish origin.

Numerous theories, besides the patriotically motivated one, have been formulated to account for the origin of the dollar mark. According to one, the mark evolved from monographic abbreviations of I H S found on some Roman coins of the time of Nero. Again, the Portuguese symbol *cifrao*, used to designate thousands, has suggested another possible origin, because the Portuguese symbol is written exactly like our dollar mark. A noted historian presented a theory concerning the origin of the dollar mark from the stamp of the mint at Potosi in Bolivia. Still another theory concerns the Spanish silver coin known as the "pillar dollar," which was widely used in the Spanish-American colonies during the seventeenth and eighteenth centuries. This coin has the Pillars of Hercules stamped on it, and it was supposed that the pillars were later copied into commercial documents. The pillar theory is strengthened because of a scroll or banner which was twisted about the pillars, and which could have suggested the S part of our dollar mark. Some theorizers have connected the dollar mark with the figure eight, and from this point of view see its origin in the Spanish dollar, which was known as a "piece of eight" because it was equivalent to eight smaller units called "reales." The aforementioned "pillar dollar" has an eight between the pillars, and the combination of the pillars and the eight has suggested to some a possible origin of our dollar mark. Consideration of the "piece of eight" also brought about, on the part of others, a theory connecting the dollar mark with a combination of the letter *p* and the numeral 8. One or more of these early theories may seem reasonable, but none was sufficiently established or authenticated by proper documentary evidence.

The American historian of mathematics, Florian Cajori, finally settled the matter of the origin of our dollar mark. With great scientific care and much documentary evidence, he traced the origin of the dollar mark to the sixteenth-century Spanish contraction pss, ps, or p^s for *pesos*. The florescent p^s evolved into $ about 1775 by English-Americans who came into business contact with Spanish-Americans. The Spanish-Americans placed the ps or p^s *after* the numerals, but the English-Americans, accustomed to writing £ *before* the number of pounds, moved the abbreviation to the left of the numerals. In the supporting documentary evidence furnished by Cajori, the modern dollar mark in the making is particularly strikingly illustrated in a letter written by Oliver Pollack to George Rogers Clark in 1778; here both the p^s and the modern dollar mark appear side by side, the latter unquestionably resulting as a simplification of the former. It was some years, in fact not until the opening of the nineteenth century, before the dollar mark appeared in print.

The word "dollar" appears to have come from the German word "thaler," a diminutive of "Joachimsthaler," a silver coin found in Joachimsthal of Bohemia about 1518.

105° *Discarded by the printer.* At one time it was suggested that the symbols ୨ and ρ be used for the constants *e* and *i*, respectively, but printers balked at making the necessary new type and so the old symbols were retained.

For a couple of other instances where mathematical notation has been affected by consideration for the printer, see Item 307° (the solidus notation for fractions) and Item 333° (the exclamation point for factorial) of *In Mathematical Circles*.

106° *Quod erat demonstrandum.* It became customary to abbreviate the clause, "(Being) what it was required to prove," which closes each proof of a theorem in Euclid's *Elements*, by the letters Q.E.D. (quod erat demonstrandum; that which was to be proved), and this time-honored procedure was carried over into many of the older high school geometry textbooks of our country. The symbol ▮, first suggested by Paul R. Halmos, or some similar sign, is frequently used today to signalize the end of a proof. Some wit has suggested using w^5, as an abbreviation of "which was what was wanted."

ARITHMETIC AND ALGEBRA

THE rest of this second quadrant will be devoted to stories and anecdotes connected with certain particular subject-matter courses of school and college mathematics. Much of the material could be classified as *classroom humor* (of varying degrees of age and quality), and it could easily be expanded into a booklet all of its own. A goodly number of the stories were sent to me by Alan Wayne, that New York City teacher *extraordinaire*, known to his many friends and admirers for his delightful humor and personality.

107° *Ethical arithmetic.* Joseph Jablonower, speaking before the Association of Teachers of Mathematics of New York City, told this story: A little girl was working at her arithmetic workbook. The sentences read: "The fox ate TWO little rabbits," "The fox ate THREE little rabbits," and then came, "The fox ate ---- little rabbits," with a picture of four rabbits. The little girl thought a while, then wrote: "The fox ate POOR little rabbits." Said Mr. Jablonower, "One should not substitute statistics for morality!"—ALAN WAYNE

108° *Thorough job.*
Teacher. Haven't you finished adding up those numbers yet?
Student. Oh, yes. I've added them up ten times already.
Teacher. Excellent! I like a student who is thorough.
Student. Thank you. Here are the ten answers.—ALAN WAYNE

109° *Living costs a century ago.* [The following is adapted, with permission, from the article of the same title, by Cecil B. Read, that appeared in the Historically Speaking section of *The Mathematics Teacher*, February, 1959, p. 124.]

The author of a textbook often finds it difficult to keep current with prices, wages, and the like. Students are quick to pick up discrepancies between values stated in problems in a textbook and actual conditions in life. Nevertheless, although there may be some lag, textbooks do offer an interesting commentary on prices of approximately the period in which they are published.

An arithmetic text of approximately a century ago was examined for some of the facts and prices which follow. The author of the text was Benjamin Greenleaf, and the title of the text is *Introduction to the National Arithmetic on the Inductive System, Combining the Analytic and and Synthetic Methods; in which the Principles of the Science are Fully Explained and Illustrated. Designed for Common Schools and Academies.* (Quite obviously the author did not believe in brief titles.) This was the "new stereotyped edition" published in Boston in 1858.

In the line of food prices, one finds problems involving milk at 5¢ a quart, raisins at 7¢ a pound, butter at 12¢ a pound, beef at 9¢ a pound, cheese at 9¢ per pound, and coffee at 25¢ a pound. Not all prices were so appreciably lower than the present-day prices. For example, lemons are quoted at 88¢ per dozen, sugar at 8¢ a pound, tea at 62¢ per pound. Among other commodities, one finds a problem involving a price of $5 for a good pair of boots, $12 for a coat, $6 for a vest, and $73 for a watch. Corn is involved in problems at a price of 75¢ per bushel, wheat at 95¢ per bushel, coal at $10 per ton. Textbooks seem considerably less expensive; a problem mentions a student paying 25¢ for an arithmetic and 67¢ for a geography.

The student of the present generation who chooses to browse through such a book may be somewhat puzzled by firkins of butter, hogsheads of molasses, quintals of fish, nails of cloth. Likewise he would probably not have much basis for comparison with the prices for wagons, harness, chaises, or yokes of oxen. He might likewise wonder what was the content of "temperance wine."

However, before one concludes that it would have been a very fine time in which to live, it would be worthwhile considering some wages.

The problems involving rates of pay are relatively few in this text, but we do find such things as: carpenters working a day fifteen hours long, a clergyman's salary of $700 a year, a laborer being paid 37¢ per day, another laborer's wages being quoted at $7 per week, farm labor at $3 per month and board, and plastering being done for 10¢ per square yard.

110° *Empty set.* On a ten-question number-completion test, Alan Wayne marked a student wrong because he did not fill in the answer to one of the questions. Indignantly, the student came up to Mr. Wayne's desk.

"Mr. Wayne," the student said, "you didn't give me credit for this answer!"

"Of course not," Mr. Wayne replied, "you did not put down the answer—zero."

"But I did so," insisted the student, "nothing was the answer, and nothing is what I have right here!"

Once more Mr. Wayne had to deliver the triply-null explanation: "Zero is not nothing!" And this occurred on a day when Mr. Wayne complained of being worn out from moving so many decimal points all day long, carrying remainders, and raising terms. He said he felt as depressed as a first-degree polynomial.

111° *Half nuts.*

Teacher. Tea costs 40 cents a pound; halibut sells for 29 cents a pound; the worst pizzas are made in Naples, Italy; ice cream originated in France. How old am I?

Student. 50 years old.

Teacher. You must have guessed. Surely you didn't figure it out logically, did you?

Student. Oh yes. You see I have an uncle who is 25 years old, and he's only half nuts.

112° *Buying grapefruit.*

"What is the price of grapefruit?"

"Two for 25 cents."

"How much is one?"

"Thirteen cents."

"Oh, then I'll take the other one."

113° *Thoroughly logical.*

Teacher. You say you got $(a + b)/(a - b)$ to reduce to 1 by cancelling. But if you cancelled the a's and the b's, wouldn't it be more logical for you to get just "+" divided by "−"?

Student. Yes sir. Then the "−" would cancel out in both places, leaving the "1".—ALAN WAYNE

114° *Modern method.* Harry Schor caught one of his algebra students counting on his fingers.

"Just what are you doing?" he demanded.

"Using a digital computer, sir," was the prompt answer.

ALAN WAYNE

115° *Of course.* During a mathematics curriculum meeting, a committee member asked, "On this page you have 'Linear Equations in One Unknown,' and on the next page you have 'Linear Equations (in Two Unknowns).' Why the parentheses in the second case?" The answer was, "Because by this time the student has learned to remove parentheses!"—ALAN WAYNE

116° *Remains to be seen.* The briefest, most tantalizing incorrect proof is that for the Remainder Theorem. Using division:

$$\begin{array}{rll} x - a) & f(x) & \quad (f \\ & \underline{f(x) - f(a)} & \\ & \qquad\quad f(a) & \end{array}$$

ALAN WAYNE

117° *Algebraic lines from Lewis Carroll.*

Yet what are all such gaieties to me
Whose thoughts are full of indices and surds?

$$x^2 + 7x + 53$$
$$= \frac{11}{3}.$$

118° *One degree more abominable.* In discussing two fellow writers, Cornelius Mathews and William Ellery Channing, Edgar Allen

Poe once wrote: "To speak algebraically: Mr. M is execrable but Mr. C is $(x + 1)$-ecrable."

119° *The Josephus problem.* According to a legend told by Hegesippus, the famous Jewish historian Flavius Josephus (ca. 37–95) once saved his own and a friend's life by a stratagem. After the Romans had captured Jotapat, the legend says that Josephus, his friend, and thirty-nine Jewish companions took refuge in a cave. The companions voted to die rather than be taken by their conquerors. Josephus and his friend, not wishing to die and yet not daring to disagree with the general vote, feigned agreement. Josephus even proposed an arrangement whereby the deaths could take place in an orderly fashion. The 41 men were to arrange themselves in a circle, then every third man was to be killed until but one was left, and he was to commit suicide. The plan was accepted. Josephus saved himself and his friend by placing himself and the friend in the 16th and 31st places.

The Josephus problem appeared in many later versions. In medieval times it concerned fifteen Turks and fifteen Christians on board a storm-tossed ship, which was certain to founder unless half the passengers were thrown overboard into the sea. After suitably arranging themselves in a circle, the Christians proposed that every ninth person, reckoned from a given starting point, be cast overboard. The Christians so disposed themselves about the circle that every infidel was eliminated and all the Christians were saved. The arrangement determined by the Christians was that shown in Figure 14, where C stands for a Christian and T for a Turk. This order can be recalled from the positions of the vowels in the mnemonic line: *From numbers' aid and art, never will fame depart*, where *a* stands for 1, *e* for 2, *i* for 3, *o* for 4, and *u* for 5. The order is thus *o* Christians, *u* Turks, *e* Christians, *a* Turks, and so on.

In other versions the people involved are Christians and Jews, scholars and dullards, whites and blacks, innocent and guilty, and so on. An ingenious soul or racist, with a knowledge of mathematics, always manages to save the "favored" group, throwing all Christian and other virtues aside in the process.

A Japanese version of the Josephus problem involves a man's thirty children, fifteen by a first marriage and fifteen by a second one.

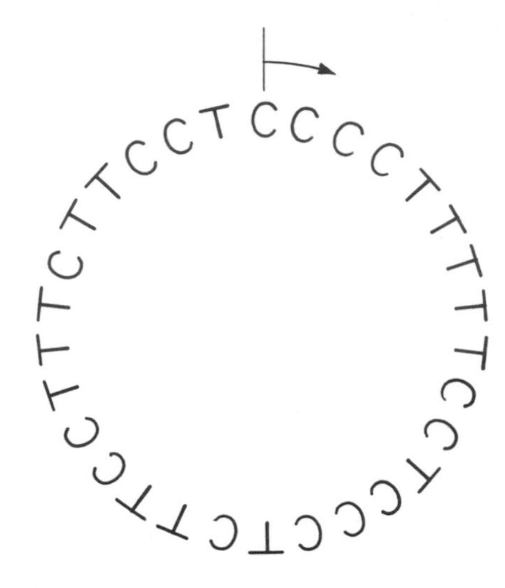

FIGURE 14

It is agreed by the father and the surviving second wife that the father's estate is too small to be divided among the thirty children, and that it should all go to just one child. The wife accordingly proposes that the thirty children be arranged in a circle and that every tenth child be eliminated until only one remains. The proposal is accepted and the wife cleverly places the children so that the fifteen children by the first marriage are first eliminated. After fourteen of the children by the first marriage have been eliminated, and the next one to go is seen to be the last child by the first marriage, the astonished father suggests that they should now start with that remaining child and reverse the direction of counting. Unable to refuse, but feeling the odds are certainly in her favor, the second wife agrees, but finds to her dismay that now all of her fifteen children are eliminated, leaving the one child of the first marriage as heir.

The general Josephus problem is amenable to simple algebraic analysis, and was first solved by P. G. Tait in 1900. Many variations of the problem have been proposed.

120° *Mathematical induction.* A man with very little hair on his head is said to be bald. Now if a man is bald, then certainly a man having only one more hair on his head is also bald. It follows by the principle of mathematical induction that all men are bald.

121° *The principle of the smallest natural number.* Every natural number is interesting. For suppose not, and let M be the set of those natural numbers which are not interesting. Then M is nonempty. Therefore, by the principle of the least natural number, M contains a smallest natural number m. That is, m is the smallest natural number that is not interesting. Ah, but that is interesting!

122° *Inverse golden rule.* One way of finding the inverse of a nonsingular matrix A is to carry A into an identity matrix I by a sequence of elementary row operations; then the same sequence of elementary row operations performed on I will carry I into the sought inverse A^{-1}. The procedure may be summarized by the *inverse golden rule*: "Do unto I as you would do unto A."

123° *The skeleton key of mathematics.* Very intriguing is the idea of a skeleton key, a key that will fit and open almost any lock. D. E. Littlewood has described modern abstract algebra as "the skeleton key of mathematics" (see his *The Skeleton Key of Mathematics: A Simple Account of Complex Algebraic Theories*, Hutchison's University Library, 1949). There is much sense in this description, for modern abstract algebra is the study of structure, and structure permeates all of mathematics. One result is that the terminology of modern abstract algebra has penetrated into practically every field of mathematics. From this point of view, modern abstract algebra has also been described as "the vocabulary of present-day mathematics."

124° *Immortality.* It has often been thought that at the moment of death one rapidly relives his entire past life. This relived life has its last moment, and this last moment has its last moment, and so on endlessly. It follows that dying may itself be eternity in that, according to the theory of limits, one approaches death but never achieves it.

GEOMETRY

125° *Presence of body.* When John Swenson was teaching a mathematics class at Wadleigh High School in New York City, he attempted to explain to a slow girl the difference between a rectangle and a rectangular solid.

"Don't you see," said Dr. Swenson, "a rectangular solid fills up space, just like a person. You have a front and a back, a top and a bottom, a right side and a left side, don't you?"

"Yes, sir!" exclaimed the student, "I'm all here!"

ALAN WAYNE

126° *Definition.*

Teacher. What is infinity?

Student. Infinity is that place which, when you reach it, is still further ahead.—ALAN WAYNE

127° *Vocational mathematics.* In a geometry class at Newton High School, Arthur Weinstein asked his class, "Name another professional worker who uses parallel lines in his work."

"A gravedigger," was the solemn answer.—ALAN WAYNE

128° *The intellectual castaway.* There was once a little group of intellectuals discussing the hypothetical situation of one of them being cast up on an uninhabited but otherwise friendly island as a lone survivor, there to remain for many years. Granting that during the castaway's stay on the island, he could have a single volume of his own choosing drift ashore, for what book should he wish, to help alleviate his loneliness and pleasantly while away the years? After some argument, it was finally unanimously decided by the group that Euclid's *Elements* might be the most consoling and most beguiling.

129° *D'Alembert on geometry.* Jean le Rond d'Alembert (1717–1783) often said things in a very apt way. Take, for example, his comment on geometry as related to physics: "Geometrical truths are in a way asymptotes to physical truths; that is to say, the latter approach the former indefinitely near without ever reaching them exactly."

130° *The Pythagorean theorem.* The following figure with its accompanying explanatory verses seems to have been devised by George Biddel Airy (1801–1892), one-time royal astronomer at Greenwich. It is quoted in Graves' *Life of Sir William Rowan Hamilton*, vol. 3 (1889).

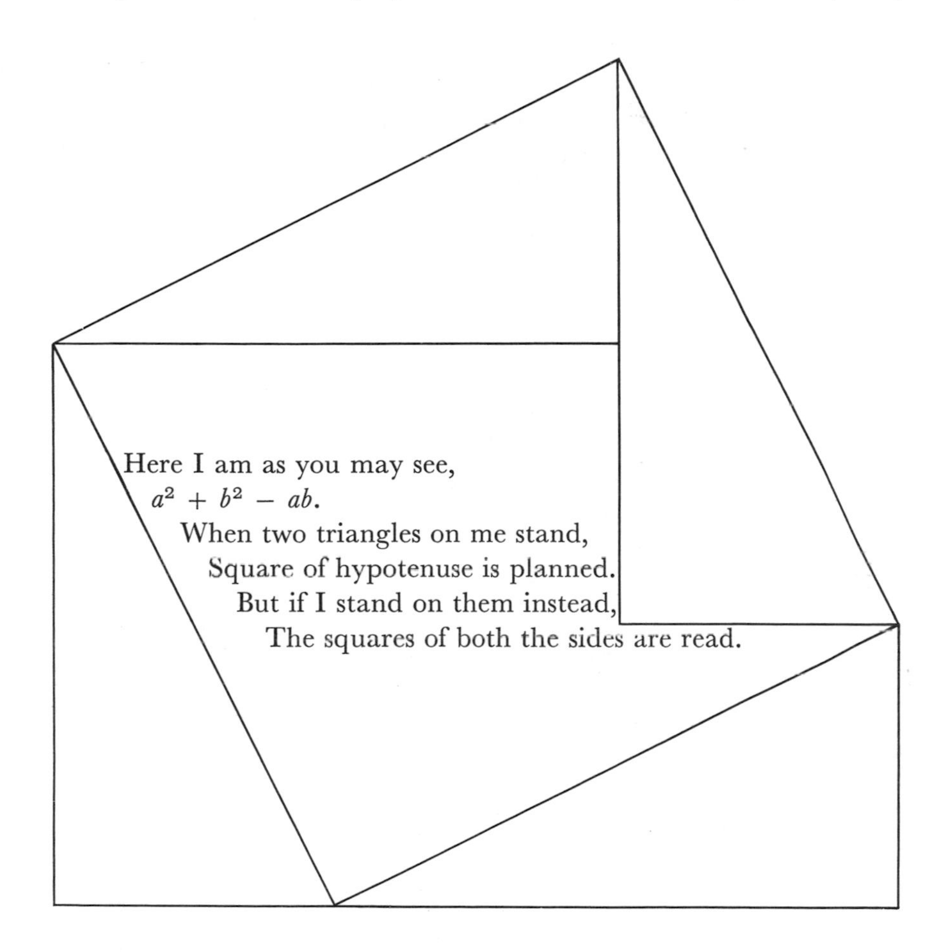

131° *The key to analytic geometry.* David Hilbert (1862–1943), in "Mathematical problems," *Bulletin of the American Mathematical Society*, vol. 8 (1902), p. 443, said: "Arithmetic symbols are written diagrams and geometrical figures are graphical formulas." And Sophie Germain (1776–1831) expressed the same thought in her *Mémoire sur les surfaces élastiques*: "Algebra is but written geometry and geometry is but figured algebra."

132° *Space communication.* With the mounting interest in space exploration and the possibility of life in other parts of the universe, suggestions have appeared from time to time concerning the construction on the earth of some enormous device that would indicate to possible outside observers that there is intelligence on our planet. The most favored device seems to be a mammoth configuration illustrating the Pythagorean theorem, built on the Sahara Desert, the Steppes of Russia, or some other vast area. All intelligent beings must be acquainted with this remarkable and certainly nontrivial theorem of Euclidean geometry, and it does seem difficult to think of a better visual device for the purpose under consideration.

133° *Where Euclid failed.* The straight line and the circle have the property of self-congruence—that is, every segment of a straight line is congruent to every other segment of the same length of the straight line, and every arc of a circle is congruent to every other arc of the same length of the circle. There is a third curve that also has this property, namely the screw or circular helix. Augustus De Morgan (1806–1871) has said that had Euclid permitted all three of these curves as instruments of geometrical construction, we would never have heard of the impossibility of trisecting an angle, of duplicating a cube, or of squaring a circle. An interested reader may care to try to justify this remark; the results should make an attractive article for a journal devoted to the teaching of mathematics.

134° *A crackpot.* Simon Newcomb (1835–1909), the famous American astronomer, tells, in his *The Reminiscences of an Astronomer* (1903), that there was a graduate of the University of Virginia who maintained that geometers are in error in assuming that lines have no thickness. Based on this view, the fellow published a school geometry text which received the endorsement of an influential New York school official, and as a consequence the book was accepted, or nearly accepted, as a textbook for the public schools of New York.

135° *Curves with proper names.* In mathematics there are many curves named after a discoverer or some subsequent researcher. Among such curves are: witch of Agnesi, spiral of Archimedes, lemniscate of Bernoulli, Bertrand curves, Bolzano curve, Bouligand curve, Bowditch curves, Cartesian ovals, Cartesian parabola, Cassini's ovals, trisectrix of Catalan, Cayley's sextic, Cornu's spiral, Cotes' spirals, folium of Descartes, cissoid of Diocles, hippopede of Eudoxus, kampyle of Eudoxus, Euler's spiral, parabolas of Fermat, hyperbolas of Fermat, spirals of Fermat, Gaussian curve, Geronon's lemniscate, Hilbert curve, quadratrix of Hippias, l'Hospital's cubic, kieroid (named after P. J. Kiernan), Koch curve, Lamé curves, Lissajou curves, trisectrix of Maclaurin, Mannheim curves, conchoid of Nicomedes, limaçon of Pascal, Peano curve, spiric of Perseus, Poinsot's spiral, Rouquet curves, pearls of Sluze, Tschirnhausen cubic, and Viviani's curve.

There are many surfaces with proper names.

136° *Inversion.* Under the transformation of circular inversion, the center of the circle of inversion maps into the single point at infinity of the inversive plane. One is reminded of Stephen Leacock's young nobleman and his explosive horsemanship: "Lord Ronald said nothing; he flung himself from the room, flung himself upon his horse and rode off madly in all directions."

137° *Catching lions.* In the 1930's a small group of mathematical wags banded together under the pseudonym E. S. Pondiczery, purportedly of the Royal Institute of Poldavia. The name and institution were chosen to fit the initials E.S.P., R.I.P. (standing for "extrasensory perception, rest in peace"), for the group had planned to write an article on extrasensory perception. The article never appeared in print.

Pondiczery's main interest lay in mathematical curiosa, and his best-known contribution to this field appeared in *The American Mathematical Monthly* (August-September, 1938, pp. 446–447) under the title, "A contribution to the mathematical theory of big game hunting."

Because of the facetious nature of this contribution, Pondiczery sought permission of editor-in-chief Elton James Moulton of the *Monthly* to use the pseudonym H. Pétard. The permission was granted, giving rise to the only known instance in mathematical literature of a paper published under a pseudonym of a pseudonym.

Pétard's paper listed sixteen ways of capturing a wild lion in the Sahara Desert. The first nine of these were catalogued as "mathematical methods," the next four as "methods from theoretical physics," and the concluding three as "methods from experimental physics." Acknowledgment of indebtedness was made to the Trivial Club of St. John's College of Cambridge, England, to the M.I.T. chapter of the Society of Useless Research, to the F. o. P. of Princeton University, and to "numerous individual contributors, known and unknown, conscious and unconscious." Here, with permission from *The American Mathematical Monthly*, we give a selection of a dozen of Pétard's methods.

1. Mathematical methods

1. THE HILBERT, OR AXIOMATIC, METHOD. We place a locked cage at a given point of the desert. We then introduce the following logical system:

AXIOM I. *The class of lions in the Sahara Desert is non-void.*
AXIOM II. *If there is a lion in the Sahara Desert, there is a lion in the cage.*
RULE OF PROCEDURE. *If p is a theorem, and "p implies q" is a theorem, then q is a theorem.*
THEOREM I. *There is a lion in the cage.*

2. THE METHOD OF INVERSIVE GEOMETRY. We place a *spherical* cage in the desert, enter it, and lock it. We perform an inversion with respect to the cage. The lion is then in the interior of the cage, and we are outside.

3. THE METHOD OF PROJECTIVE GEOMETRY. Without loss of generality, we may regard the Sahara Desert as a plane. Project the plane into a line, and then project the line into an interior point of the cage. The lion is projected into the same point.

4. THE BOLZANO-WEIERSTRASS METHOD. Bisect the desert by a

line running N–S. The lion is either in the E portion or in the W portion; let us suppose him to be in the W portion. Bisect this portion by a line running E–W. The lion is either in the N portion or in the S portion; let us suppose him to be in the N portion. We continue this process indefinitely, constructing a sufficiently strong fence about the chosen portion at each step. The diameter of the chosen portions approaches zero, so that the lion is ultimately surrounded by a fence of arbitrarily small perimeter.

5. THE "MENGENTHEORETISCH" METHOD. We observe that the desert is a separable space. It therefore contains an enumerable dense set of points, from which can be extracted a sequence having the lion as limit. We then approach the lion stealthily along this sequence, bearing with us suitable equipment.

6. THE PEANO METHOD. Construct, by standard methods, a continuous curve passing through every point of the desert. It has been remarked that it is possible to traverse such a curve in an arbitrarily short time. Armed with a spear, we traverse the curve in a time shorter than that in which a lion can move his own length.

7. A TOPOLOGICAL METHOD. We observe that a lion has at least the connectivity of the torus. We transport the desert into four-space. It is then possible to carry out such a deformation that the lion can be returned to three-space in a knotted condition. He is then helpless.

2. Methods from theoretical physics

10. THE DIRAC METHOD. We observe that wild lions are, *ipso facto*, not observable in the Sahara Desert. Consequently, if there are any lions in the Sahara, they are tame. The capture of a tame lion may be left as an exercise for the reader.

11. THE SCHRÖDINGER METHOD. At any given moment there is a positive probability that there is a lion in the cage. Sit down and wait.

13. A RELATIVISTIC METHOD. We distribute about the desert lion bait containing large portions of the Companion of Sirius. When enough bait has been taken, we project a beam of light across the

desert. This will bend right round the lion, who will then become so dizzy that he can be approached with impunity.

3. Methods from experimental physics

14. THE THERMODYNAMICAL METHOD. We construct a semi-permeable membrane, permeable to everything except lions, and sweep it across the desert.

15. THE ATOM-SPLITTING METHOD. We irradiate the desert with slow neutrons. The lion becomes radioactive, and a process of disintegration sets in. When the decay has proceeded sufficiently far, he will become incapable of showing fight.

138° *The oui-ja board curve.* One of the most interesting mathematical museum pieces might have been a certain oui-ja board, or more properly speaking, planchette,* used in England sometime in the late nineteenth century. Since the controlling spirit of the planchette was believed to be a one-time senior wrangler, two young ladies were led to ask the planchette to write the equation of its own bounding curve. Three or four times the board quite distinctly gave the polar equation

$$r\theta = a \sin \theta.$$

Because of an incorrect plotting of the locus of this equation, the young ladies felt the board must be fibbing. However, when their mathematics instructor correctly plotted the locus, the ladies were surprised to see the remarkable resemblance between the locus and the bounding curve of the planchette.†

The locus appears as in Figure 15. It is symmetric in the polar axis, and in each half of the plane there is an infinite set of nesting

* A *planchette* is a small board on two castors and a vertical pencil, which purportedly *writes* answers to questions without the conscious effort of the questioners, whose fingers rest lightly on the board. A *oui-ja board* is a small board on legs resting on a larger board marked with the letters of the alphabet, and it purportedly *spells out* answers to questions, without the conscious effort of the questioners, whose fingers rest lightly on it, by successively touching the appropriate letters on the larger board.

† The planchette incident is rather completely written up in Sir Oliver Lodge, *The Survival of Man*, pp. 130–134.

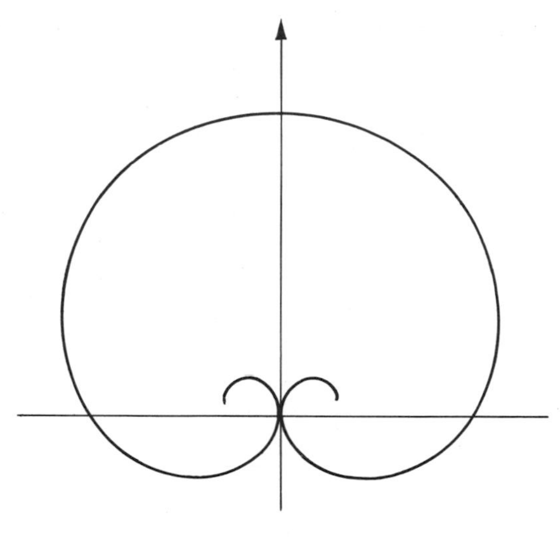

FIGURE 15

ovals, all tangent to the polar axis at the pole. The intercept on the polar axis is the constant a.

Since this curve enjoys the probably unique distinction of having been given by the oui-ja board, and because the only other name ever given to the curve, the *cochlioid*, has also been given to other curves, it has been proposed to name the curve the *oui-ja board curve*.

The oui-ja board curve possesses a number of interesting geometrical properties. It can be shown that with a celluloid template of the form of half the oui-ja board curve one can multisect angles, rectify circular arcs, and locate centroids of circular arcs and circular sectors. Such a template would make a useful graphometer or draftsman's tool. The curve seems to have been overlooked by writers on graphical calculus. In certain aspects it seems superior to the similarly used spiral of Archimedes, hyperbolic spiral, and circular involute.*

139° *Anna.* Bowdoin College, located at Brunswick, Maine, is the oldest institution of higher learning in the state; it was incorporated in 1794, when Maine was still a part of Massachusetts, and it opened in

* For the mathematics and applications of this template see Howard Eves, "A graphometer," *The Mathematics Teacher*, November, 1948, pp. 311–313.

1802. It is a highly reputable college, possessing a long list of famous graduates, such as Henry W. Longfellow, Nathaniel Hawthorne, Robert E. Peary, Franklin Pierce, Thomas B. Reed, and Chief Justice Melville W. Fuller. Being such an old college, the campus exhibits an interesting mixture of the modern and the venerable.

Among the more patriarchal buildings is Appleton dormitory, and an observant visitor will discover, just to the right of the entrance at the south end of the building, a small memorial stone set low to the ground and bearing the simple inscription

ANNA '78

It is likely that no amount of hypothesizing on the part of the visitor will yield the correct explanation to account for this memorial stone. Actually, it is one of the legends of the institution. It seems that it was back in 1878 that analytic geometry was first introduced as a course of study at the college. The subject proved to be highly unpopular, and at the end of the school year the students consigned their analytic geometry textbooks to a bonfire, buried the ashes close to the south end of the dormitory, and set a small stone memorial marker in the ground to commemorate the event.

One cannot but wonder why the subject of analytic geometry proved so unpopular, for surely, if it is taught at all well, there are few academic experiences that can be more thrilling to the student of elementary college mathematics than his introduction to this powerful and productive *method* of attacking geometrical problems. The student of mathematics meets, for the first time and on a truly colossal scale, the great *transform-solve-invert* technique of mathematics, whereby a problem which is found to be difficult in one field of mathematics is cleverly shifted to a corresponding problem in another field of mathematics in which the student may be more facile. For example, if one were asked to find the Roman numeral representing the product of the two given Roman numerals LXIII and XXIV, one would *transform* the two given Roman numerals into the corresponding Hindu-Arabic numerals, 63 and 24, *solve* the related problem in the Hindu-Arabic notation by means of the familiar multiplication algorithm to obtain the product 1512, then *invert* this last into Roman notation, finally obtaining MDXII as the answer to the original problem. By an appropriate transformation, a difficult problem has been converted into an easy problem. Such is the *method* called analytic geometry, which ingeniously transforms a geometrical problem into an associated algebraic one.

140° *Paul Kelly on non-Euclidean geometry.* The delightful and eminent American geometer, Paul Kelly, experimented during the 1969 fall semester with a mathematics appreciation course at the University of California in Santa Barbara. The course was open to the general college student, and it was Professor Kelly's hope in the course to relate various mathematical ideas to the real world of the student. For example, he thought it would be interesting to the student to note a possibly important implication of non-Euclidean geometry for understanding oneself. His thesis ran as follows:

From early school days we get used to the Euclidean explanation of space, and for ordinary problems this system of ideas works—indeed, it works so satisfactorily that we regard it as the "true" system, and we tend to feel an ingrained disbelief in non-Euclidean geometry. Now each person is a kind of system of beliefs, emotions, attitudes, and so forth, picked up from his parents, teacher, society, and experiences in general, and for him, in ordinary circumstances, this system "works." As a result he tends to regard his system as "true," and to view with suspicion, perhaps even with contempt or disbelief, any personal philosophy, religion, or culture that differs from his own, such as those of foreigners or of people in a different stratum of society. Achieving a genuine understanding of someone radically different from oneself is thus seen to be emotionally parallel to achieving a genuine understanding of the possibility of our world's being closer to some non-Euclidean model than to a Euclidean one.

TRIGONOMETRY

141° *Striking out.* The latest exploit of Hapless Harry, who, unlike Lucky Larry, just can't get the right answer, was to cancel the n's to get:

$$y = \sin \frac{x}{n} \quad \text{becomes} \quad y = \text{six}.$$

ALAN WAYNE

142° *Functional grammar.* In the Freilich-Shanholt-McCormack *Trigonometry*, there occurs a "triple negative." The question is asked: "Which of the trigonometric functions is never negative?" The answer is given: "None is never negative."—ALAN WAYNE

143° *A Clayton Dodge whimsy.* On page 22, where circular coordinates and their notation are introduced, of Clayton Dodge's *The Circular Functions*, appears the footnote: "Since it is customary to use *round* parentheses to enclose *rectangular* coordinates, it seems equally logical to use *square* brackets to enclose *circular* coordinates."

144° *Pie in the bathroom.* The digits in the decimal expansion of π seem to satisfy no pattern and appear to occur in an absolutely random fashion. Because of this, the successive digits in the decimal expansion of π have been used where a random sequence of the digits is required. Use of this randomness can be made in laying tile, where, to break the monotony of a uniform color, one wishes every now and then to insert a tile of a different color. Clayton Dodge employed this idea when tiling a wall of his bathroom in basic white tiles interspersed randomly with tan ones.

There is an exterior wall of the large Upton-Hastings dormitory at Gorham State College in Gorham, Maine, which is divided into panels that are tiled in "random" fashion. How very interesting it could have been if one panel utilized the decimal expansion of π, another the decimal expansion of e, another the decimal expansion of $\sqrt{2}$, another that of log 2, still another that of Euler's constant γ, and so on. A panel of this sort, employing the decimal expansion of π, is shown in the frontispiece of this volume.

145° *The happiest mnemonic in elementary mathematics.* Although the word *mnemonic* strictly is an adjective, meaning "assisting the memory," it is frequently incorrectly used as a noun, with the sense of "any device or system assisting the memory." In this latter sense, there are a number of mnemonics in elementary mathematics. For instance, in Item 41° we considered some mnemonics for recalling π to a large number of decimal places. In an elementary trigonometry class, the mnemonic "All Students Take Calculus" is used to recall that in the first quadrant All the trigonometric functions are positive; in the second quadrant only the Sine and its reciprocal, the cosecant, are positive; in the third quadrant it is only the Tangent and its reciprocal, the cotangent; in the fourth quadrant it is only the Cosine and its reciprocal, the secant.

The finest mnemonic in elementary mathematics is the so-called *rules of circular parts*, given by John Napier for the easy recall of ten formulas which are useful in solving spherical right triangles. In Figure 16 is pictured a spherical right triangle lettered in conventional

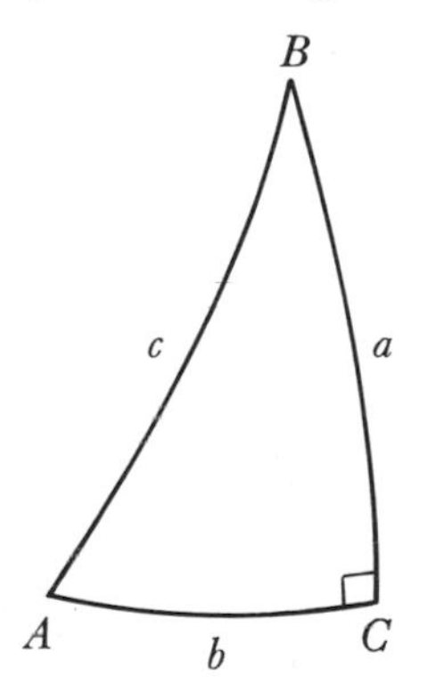

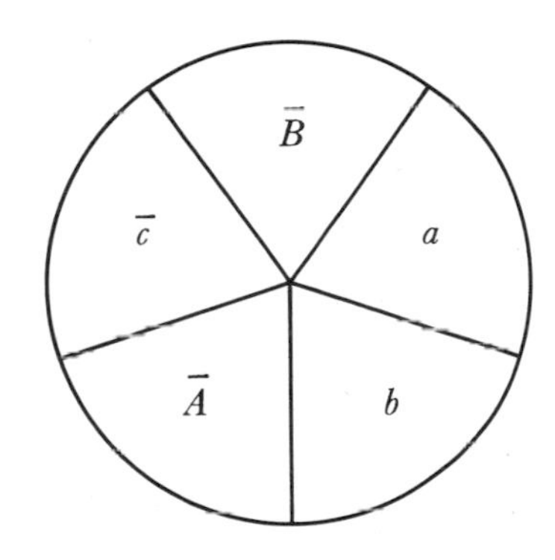

FIGURE 16

manner. To the right of the triangle appears a circle divided into five parts, containing the same letters as the triangle, except C, arranged in the same cyclic order. The bars on c, B, A mean *the complement of* (thus $\bar{B}$ means $90° - B$). The angular quantities a, b, $\bar{c}$, $\bar{A}$, $\bar{B}$ are called the *circular parts*. In the circle there arc two circular parts contiguous to any given part, and two parts not contiguous to it. Let us call the given part the *middle part*, the two contiguous parts the *adjacent parts*, and the two noncontiguous parts the *opposite parts*. Napier's rules may be stated as follows:

1. The s*i*ne of any m*i*ddle part is equal to the product of the c*o*sines of the two *o*pp*o*site parts.
2. The s*i*ne of any m*i*ddle part is equal to the product of the t*a*ngents of the two *a*dj*a*cent parts.

By applying each of these rules to each of the circular parts, one obtains the ten formulas used for solving spherical right triangles.

146° *A knowledge of trigonometry.* J. L. Walsh once remarked in connection with Fejér's theorem on Fourier series that the reason that Fejér was able to make such significant contributions to trigonometric series was that he knew so much trigonometry. I am fond of extending this by claiming that the reason Dirichlet was able to prove something about convergence of Fourier series whereas Fourier was not was that Dirichlet knew more trigonometry.—RALPH P. BOAS

PROBABILITY AND STATISTICS

147° *Significant figures.* A touring lecturer started off his favorite lecture: "A million years ago this earth was trod by dinosaurs." He was immediately interrupted by a well-meaning old lady in the audience, who said, "You mean a million and eight years ago, don't you?" "Why do you say that?" queried the lecturer. "Because I heard you give this same lecture eight years ago," explained the old lady.

148° *The breathy baboon.*

There once was a breathy baboon
Who always breathed down a bassoon,
For he said, "It appears
That in billions of years
I shall certainly hit on a tune."
SIR ARTHUR EDDINGTON

149° *A remarkable probability.* A teacher was discussing probability and showed the probability that the event must happen is 1. He then stated *emphatically:* "The probability of the event not happening is 0!" (0! = 1.)—VERNER E. HOGGATT, JR.

150° *Unlikely.* A new mother was proudly showing off her twins to a friend. "Do you know," she boasted, "twins happen only once in 860 times!"

"Is that so?" said her friend, "I wonder how you found time for the housework!"—ALAN WAYNE

151° *Choice, not chance.* A German professor of statistics complained to his baker that the loaf of bread he bought each day was lighter than the stated weight. After that, as the professor found by weighing, the loaf of bread he received was either equal to or more than the stated weight. After a month of this, the professor reported the baker to the police as one who was deliberately short-weighing his customers. "Since I am not receiving a normal distribution of weights," he said, "obviously I am getting the heavier 'tail-end' of the distribution, so that almost everyone else is being given short weight!"

ALAN WAYNE

152° *To be fair.* Before tossing the coin the student said, "If it comes up heads, I shall go to the movies; if it comes up tails, I shall go for a beer; if it balances on edge, I shall study."

153° *Something certain about the probable.* Descartes, in his *Discourse on Method,* says: "It is a truth very certain that, when it is not in our power to determine what is true, we ought to follow what is most probable."

154° *Faith in a traditional event.* Tradition is the handing down of statements, beliefs, legends, customs, etc., from generation to generation, especially by word of mouth or by practice. Now credence in a past event supported solely by tradition tends to diminish as the length of the tradition increases, and efforts have been made to formulate an appropriate governing law. Perhaps the most famous solution to the problem is that given in Craig's *Theologiae Christianae Principia Mathematica*, published in 1699. There it is maintained that credence in a traditional event varies inversely as the square of the time from the beginning of the tradition. On this principal, it is concluded that faith in the Gospel, so far as it depends on oral tradition, essentially expired about the year 880, and that so far as it depends on written tradition it will essentially expire about the year 3150. Peterson, by adopting a different law of diminution, concluded that faith would essentially expire in 1789.

De Morgan, in his *Budget of Paradoxes*, quotes the Cambridge orientalist Samuel Lee to the effect that Mohammedan writers, replying to the argument that the Koran does not possess the evidence derived from Christian miracles, contend that as faith in the evidence of the Christian miracles daily weakens, a time will come when that faith will fail to afford assurance that they were miracles at all. Then there will be need for another prophet and other miracles.

Every religion resting on tradition will in time wither away.

155° *Specious statistics.* Statistics show that over the years our principal roads have been made wider and wider, and at the same time accidents have increased. Apparently wider roads (perhaps because they cause people to drive faster) are a cause of accidents.

156° *Statistical prediction.* NEW AUTOS WILL HIT SIX MILLION.

157° *A resolution.* Statistics reveal that our cars on the road kill 100 people daily. Let's resolve to do better.

158° *Volvos.* "Nine out of ten Volvos registered in the United States in the past eleven years are still on the road." (Nine out of eleven Volvos registered here in the past ten years were probably registered in the past *four* years!)—CLAIRE RUBIN

159° *Degrees of classification.* There are liars, there are damn liars, and there are statisticians.

160° *Railroad statistics.* A Pullman porter was asked by a passenger, "What's the average tip you get from people?"

"Five dollars, sir," replied the porter.

The passenger handed the porter five dollars.

"Mister," exclaimed the porter, "you're the most average person I've met!"—ALAN WAYNE

161° *Enumeration.* A great danger in enumeration problems lies in unwittingly duplicating one's counting. As a gross illustration consider the following question: "How many legs does a cow have?"

"Well, let's see. There are two in front, two in back, two on each side, and—oh, yes, one in each corner. That makes twelve in all, sir."

"What about the cow's forelegs?"

162° *Extrapolation.* A motor-vehicle bureau survey has revealed that, in 1940, each car on the road contained an average of 3.2 persons. In 1950, occupancy had declined to 2.1 persons per car. By 1960, the average was down to 1.4 persons. If we project the statistics to 1980, every third car going by will have nobody in it.

LOGIC

163° *Paradoxes.* Perhaps the greatest paradox is that there are paradoxes in mathematics.

164° *The antinomies of logic.* When, in a logic course, one is discussing the ancient paradoxes of Eubulides and Epimenides, it seems an appropriate time to tell the story of the Italian tailor. It seems that a chap accidentally almost tore off a patch pocket of his jacket. Taking the jacket to his tailor, he asked if it could be repaired. After looking things over, the tailor replied, "Euripides, Imenides."

165° *Logic.*

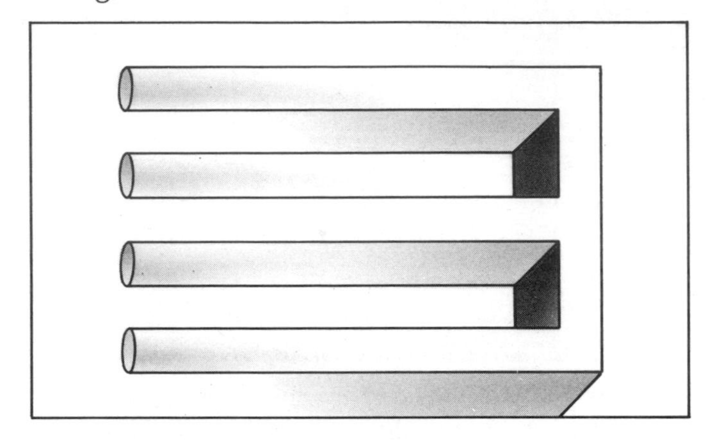

FIGURE 17

~~All is not what it seems to be.~~
~~Nothing is what it seems to be.~~
~~Nothing seems to be what it is.~~
~~Everything seems to be other than what it is.~~
~~Everything is not what it seems to be.~~
~~Not everything is what it seems to be.~~
Some things are not what they seem to be.
CLAYTON W. DODGE, *Numbers & Mathematics**

166° *Logic for the illogical.* How are you going to teach logic in a world where everybody talks about the sun setting, when it's really the horizon rising?—CAL CRAIG

167° *Proof by logic.* One is reminded of the proof by "logic," in *Robin Hood* (Chapter 14, "Robin Meets Friar Tuck"), that *every* river has only one bank:

"Well then, good fellow, holy father, or whatever you are," said

*Prindle, Weber & Schmidt, Inc., 1969.

Robin, "I would like to know if this same friar lives on this side of the river or the other."

"Truly, the river has no side but the other," said the friar.

"How can you prove that?" asked Robin.

"Why, this way," said the friar, noting the points on his fingers. "The other side of the river is the other, right?"

"Yes, that's true."

"Yet the other side is but one side, do you see?"

"Nobody could say it isn't," said Robin.

"Then if the other side is one side, this side is the other side. But the other side is the other side; therefore both sides of the river are the other side, just as I said."

"That is a good argument," said Robin. "Yet I still do not know whether the friar I am seeking is on the side of the river where we are standing or on the side where we are not."

"That," said the friar, "is a question which the rules of argument can't answer. I advise you to find out by the use of your senses such as sight, feeling, and such things."—MAURICE LAPMAN, *Robin Hood**

168° *An invalid proof.* There is a well known little problem that runs as follows: "Some sportsmen, having pitched camp, set out to go bear hunting. They walked 15 miles due south, then 15 miles due east, where they shot a bear. Walking 15 miles due north, they returned to their camp. What was the color of the bear?"

The answer is, "White," and the reasoning given is thus. In order to walk due south 15 miles, then due east 15 miles, and then due north 15 miles back to the starting point, one must start at the north pole. Any bears in the vicinity of the north pole are polar bears, and polar bears are white.

Though the above *answer* may be correct, the reasoning leading to it is incomplete, for there are other places on the earth from which a trip as described above can be made. Consider, for example, any point 15 miles north of the small southern latitude circle that has a circumference of 15 miles. Starting at this point, a walk of 15 miles due

*Globe Book Company, 1952.

south would bring one to the small latitude circle, a walk of 15 miles would take one once around this circle of latitude, and a final walk of 15 miles due north would bring one back to the starting point. One could also start at any point 15 miles north of a southern latitude circle having a circumference of 7.5 miles, or of one having a circumference of 5 miles, or, in general, of one having a circumference of $15/n$ miles, where n is any positive integer. These various starting points themselves lie on certain circles of latitude (see Figure 18). Thus the locus of

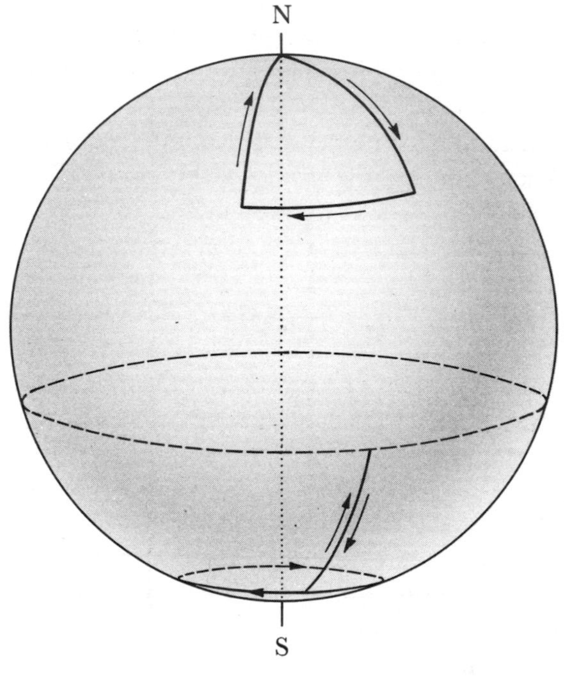

FIGURE 18

possible starting points consists of the north pole and an infinite set of small circles of latitude down near the south pole. But there are no bears in the south polar region. It follows that the hunters must have started from the north pole, and, as before, the color of the bear was white.

169° *How Abraham Lincoln improved his power of logic.* "He studied and nearly mastered the six books of Euclid since he was a member of Congress.

"He began a course of rigid mental discipline with the intent to improve his faculties, especially his powers of logic and language. Hence his fondness for Euclid, which he carried with him on the circuit till he could demonstrate with ease all the propositions in the six books; often studying far into the night, with a candle near his pillow, while his fellow-lawyers, half a dozen in a room, filled the air with interminable snoring."

ABRAHAM LINCOLN (*Short Autobiography*, 1860)

170° *Bertrand Russell "proves" he is the Pope.* It is usually taken as a principle in formal logic that the acceptance of any one logically false statement as true would permit one to *prove* any other statement whatever. When Bertrand Russell was teaching logic in the City College of New York many years ago, he stated this in class one day, and a student challenged it.

"All right," the student said, "Suppose that $1 + 1 = 1$. Prove that you are the Pope."

Russell is said to have replied without hesitation. "I am 1 and the Pope is 1, and if we are considered together we are $1 + 1$—which is 1, so we are one person and I am the Pope." This was said, according to legend, with a twinkle in his eye.

TOPOLOGY

171° *Bilateral and unilateral surfaces.* A circular disk is a bilateral surface with one unknotted edge; a Möbius band is a unilateral surface with one unknotted edge. Do there exist, in ordinary three-dimensional space, bilateral and unilateral surfaces each having one *knotted* edge? The answer is affirmative; Figure 19 pictures the former (which was discovered by F. Frankl and L. S. Pontryagin in 1930), and Figure 20 pictures the latter.

The interested reader might care to try to construct, in ordinary three-dimensional space, two-edged bilateral surfaces and two-edged unilateral surfaces where:

1. the two edges are unknotted and unlinked,

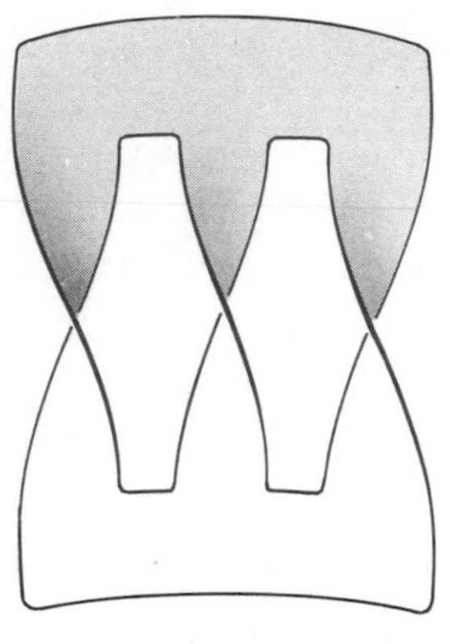

FIGURE 19

FIGURE 20

2. the two edges are unknotted and linked,
3. the two edges are knotted and unlinked,
4. the two edges are knotted and linked,
5. one edge is unknotted and one knotted, and they are unlinked,
6. one edge is unknotted and one knotted, and they are linked.

Solutions may be found in Martin Gardner, *The (First) Scientific American Book of Mathematical Puzzles and Diversions*, Simon and Schuster, 1959, pages 63–72.

172° *The Möbius strip.*

A mathematician confided
That a Möbius strip is one-sided,
And you'll get quite a laugh
If you cut one in half,
For it stays in one piece when divided.

ANONYMOUS

173° *The burleycue dancer.*

A burleycue dancer, a pip
Named Virginia, could peel in a zip;
But she read science fiction
And died of constriction
Attempting a Möbius strip.

CYRIL KORNBLUTH

174° *A river with only one bank.* If the surface of a continent assumes the form of a Möbius band and is traversed by a river following the median line of the band, one can walk from one bank of the river to the opposite bank without crossing the river; that is, the river has only one bank. (Also see Item 168°.)

175° *The witch doctor.* Consider the wisdom in the procedure of the witch doctor who gave advice to couples wondering if they should get married. If he wished to prophesy a future break-up in the proposed marriage, he would split an untwisted band; if he wished to prophesy that the couple would quarrel but still stay together, he would split a band having a full twist; if he wished to prophesy a perfect marriage, he would split a Möbius band.

176° *The frustrated painters.* Consider the plight of two painters on opposite faces of a Möbius band, one painting with white paint and the other with black paint.

177° *Klein's bottle.*

A mathematician named Klein
Thought the Möbius strip was divine.
Said he, "If you glue
The edges of two
You'll get a weird bottle like mine."

ANONYMOUS

178° *College daze.* A University of Virginia mathematics professor was amazed, when he arrived at a neighboring woman's college to lecture on "Convex Sets and Inequalities," to find the auditorium filled to capacity. A glance at the campus newspaper gave him the answer. His subject was announced as "Convicts, Sex and Inequalities."

KARL B. KNUST, JR.

179° *Jordan's theorem.* For many centuries geometers accepted as intuitively obvious the fact that a simple closed curve C in a plane p divides the rest of p into two sets in such a way that any two points in the same set can be joined by a polygonal path in p not intersecting C, whereas two points in different sets cannot be so joined. This fact certainly does seem obvious if C is a circle or a convex planar polygon, but Figures 21 and 22 show that the fact is not so obvious for more complicated simple closed planar curves, and, of course, it is possible to draw much more involved simple closed curves than those of Figures 21 and 22.

FIGURE 21

It was not until the second half of the last century that it was thought that a proof of the concerned fact was necessary. Surprisingly, such a proof was not easily found and the problem challenged mathematicians for a number of years. The first successful proof was given by Camille Jordan in his famous *Cours d'analyse*, which accounts for the connection of his name with the theorem. Though today many proofs exist, and some recent ones are considerably simpler than Jordan's original proof, they are still quite complex. If we restrict

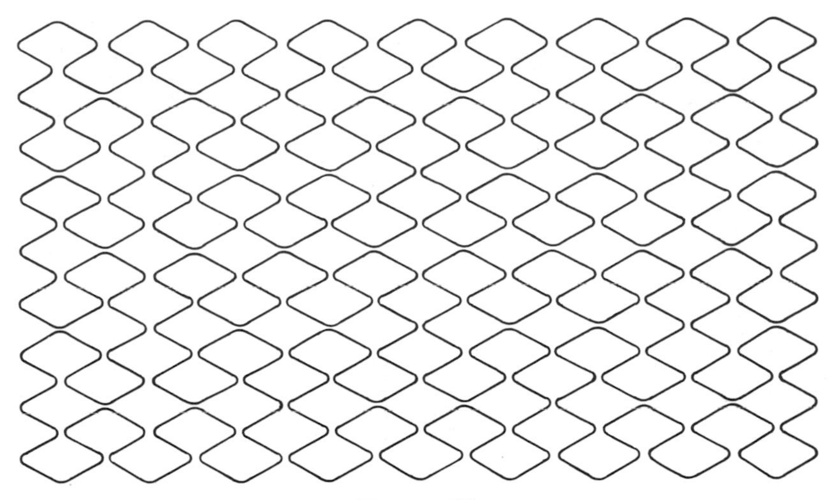

FIGURE 22

ourselves to simple closed *polygonal* curves in the plane, a relatively easy solution is possible.

The problem of recognizing the "inside" and the "outside" of a simple closed planar curve recalls the story of the drunk man observed one day walking round and round a great stone pillar of a large building and moaning to himself. When asked by a passerby why he was so sad, he replied, "Look, I'm walled in."

180° *Protestant curves.* A Jordan curve is a planar curve that does not intersect itself. Professor Lee Swinford, of the University of Maine, calls such curves "Protestant curves," because "they do not cross themselves."

QUADRANT THREE

From a Darrow cartoon
to a well integrated person

FROM THE YOUNGER SET

A second-grade teacher once confessed that though the trials and tribulations of teaching young children were very great, these were somewhat offset by the instances of humor either observed directly in the classroom or reported by parents as occurring at home. It seems that the "new math" has created many examples of the latter type.

181° *A cartoon by Whitney Darrow, Jr.* It is a family scene. The father, mother, and small son are seated at the supper table. The perplexed father has just asked his son to explain the "new math" he has been hearing so much about. The little boy replies: "Dad, you can't expect to pick up the basics of new math in a simple dinner-table conversation."

182° *A cartoon by Alan Dunn.* A confused and by now irritated father sits at his desk, which is piled high with papers and calculations, trying to work out his federal income tax. His little daughter comes in, takes up the income tax form and says: "You see, Daddy, this set equals all the dollars you earned; your expenses are a subset within it. A subset of *that* is your deductions."

183° *The aftermath of the new math.* Teaching experience indicates the value of incorporating much of the so-called "new math" in our elementary school programs, but it also advises against overdoing it. One is reminded of two schoolboy definitions for salt: (1) Salt is what, if you spill a cupful into the soup, spoils the soup. (2) Salt is what spoils your soup when you don't have any in it.

184° *Cuisinaire rods.* Anne entered first grade last year, and soon came home telling everyone about the "new math." She brought home colored blocks of graduated sizes and practiced her math with great enthusiasm. Red was twice as long as white, green three times as long, purple four times, etc. When Anne was asked what two plus three is, she replied, "Two plus three is yellow."

185° *Boners and bloopers, in twenty-five minute doses.* George Q. Lewis, Director of Humor Societies of America, has defined a *boner* as "a short and pointed mistake that produces an amusing effect," and a *blooper* as "a kingsize boner." For over a quarter of a century the Association for the Elimination of Boners and Bloopers, "a group of goops who enjoy goofery," has been submitting boners and bloopers to the Library of American Wit and Humor. More than 10,000 of these items are now in the Library, where they have been catalogued under a number of different headings by Leopold Fechtner, Curator of the Library. Boners and bloopers are continually being made in the mathematics classrooms of the country, and thence circulated among the mathematical fraternity by the teachers; here are some that have become common property. Almost any teacher of mathematics can easily add further items gleaned from his or her personal classroom experience.

1′ Trigonometry is when a man marries three wives at the same time.

2′ Algebra is an older form of arithmetic.

3′ Geometry teaches us how to bisex angels.

4′ A polygon with seven sides is called a hooligan.

5′ Algebraic symbols are used when you do not know what you are talking about.

6′ A circle is a line which meets its other end without ending.

7′ A circumference runs around outside a circle trying to get in.

8′ Squares are circles with corners.

9′ A rectangle is a sloppy square.

10′ Vertical is just the same as horizontal, only just the opposite.

11′ Arabia gave us the dismal system which we still use in counting.

12′ The Egyptian pyramid was made in the shape of a huge triangular cube.

13′ Infinity is a place you can't get to where lines meet.

14′ The earth is round but flattened at the corners.

15′ The center of most of our rugs is Persia.

16′ An average man is someone who isn't older than anyone else.

17′ Use our easy credit plan—100 per cent down, nothing to pay each month!

18′ The equator is a menagerie lion running around the earth.

19′ Sam Smith, who has been ill with arithmetic for the past few months, was able to leave the house Sunday.

20′ An Architect, serving a prison term, was heard to complain that the prison walls were not built to scale.

21′ A Scotland Yard measures two feet and ten inches.

22′ According to a recent Census Bureau report, the average American is growing older.

23′ The earth makes a resolution every twenty-four hours.

24′ When a ship passes the 180th degree of longitude, Saturday becomes Sunday.

25′ Two planes meeting at the same altitude have one end in common.

CLASSROOM TACTICS AND ANTICS

SOMETIMES an instructor must be more or less subtle.

186° *A penetrating remark.* One day a visitor to Alan Wayne's mathematics classroom found the students at the blackboard working simple problems involving real, practical situations. "What are you doing?" asked the visitor. "I'm drilling through the concrete," replied Alan Wayne.

187° *An empty stomach.* A student, looking for an excuse not to work, told his teacher that he was hungry and asked the teacher if it was wise to do the mathematics problems on an empty stomach. "It's all right," replied the teacher, "but it would be better to use a piece of paper."

188° *Sneezing.* A student very audibly sneezed during a mathematics lecture. The teacher immediately stopped his lecture and interjected: "Sneezing is a natural provision whereby the over-profound thinker might expel superfluous ideas through the nose"—a line found in Edgar Allen Poe's essay *Eureka.*

189° *A pointed remark.* A mathematics student asked a particularly stupid question in class. Not wishing to discourage the student and yet desiring to be honest, the teacher looked directly at the top of the student's head and said, "You know, you have a point there."

190° *Advice prior to giving a mathematics test.*

> We have two ends with a common link;
> With one we sit, with one we think.
> Success depends on which we use:
> Heads we win, tails we lose.
>
> ANONYMOUS

191° *Always read your problem carefully.* Every now and then a mathematics teacher must emphasize to his class the care that should be taken in reading a problem, so that the solver will be certain of exactly what is given in the problem and what is required. This lesson can be effectively accomplished by posing to the class the following teaser: "We are given an ordinary 8×8 chess board with a rook (castle) in the upper left square of the board. The problem is to move the rook, by rook moves alone*, so that it will enter each square of the board once and only once and end up in the diagonally opposite (that is, lower right) square of the board."

After the students have experimented unsuccessfully with the

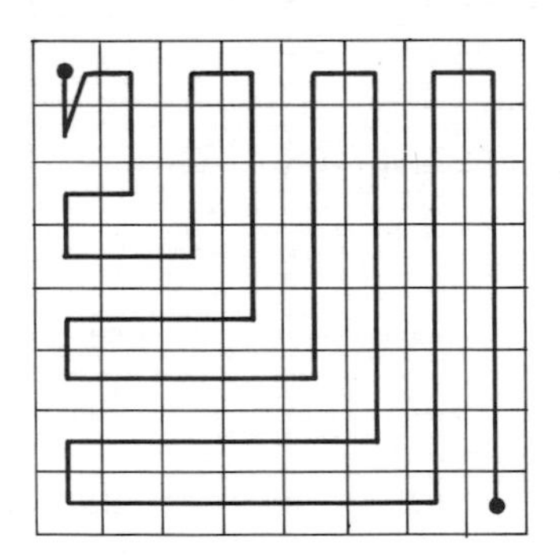

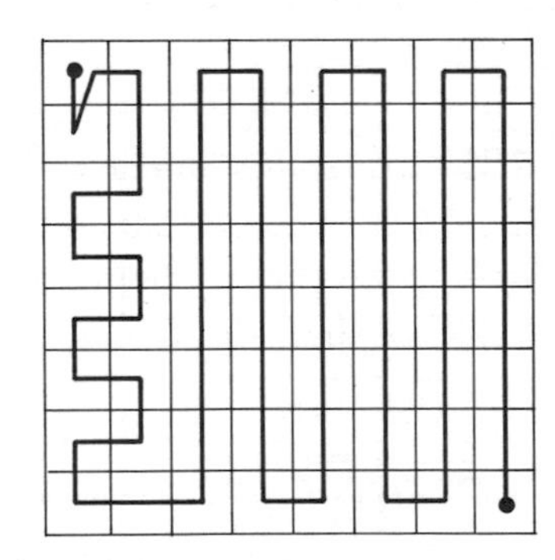

FIGURE 23

* A rook may move only along the rows and the columns of the board.

problem for a time, the teacher can exhibit the solution indicated in Figure 23, and point out the care needed in reading a problem. Since the rook was given as already in the top left square, we have not yet caused the rook to *enter* this square, and the problem says that the rook is to enter *each* square of the board once and only once.

Similar, but simpler, problems suitable for elementary school classes are:

1. I have 60 cents made up of two coins. One of the coins is not a dime. What are the two coins?
2. A farmer has three piles of hay in one field and four piles of hay in another field. How many piles will he have if he puts them all together?
3. Which is heavier, a pound of iron or a pound of feathers?
4. How much dirt is there in a hole 2 feet wide, 3 feet long, and 4 feet deep?

192° *Misdirection.* In *Death from a Top Hat*, by Clayton Rawson, the magician Merlini wishes to point out to the police how a conjurer will focus an audience's attention on the right hand so that everyone completely fails to see what the left hand is doing. He draws the diagram of Figure 24. "X is the center of the circle; BC is $9\frac{1}{2}$ inches long, and

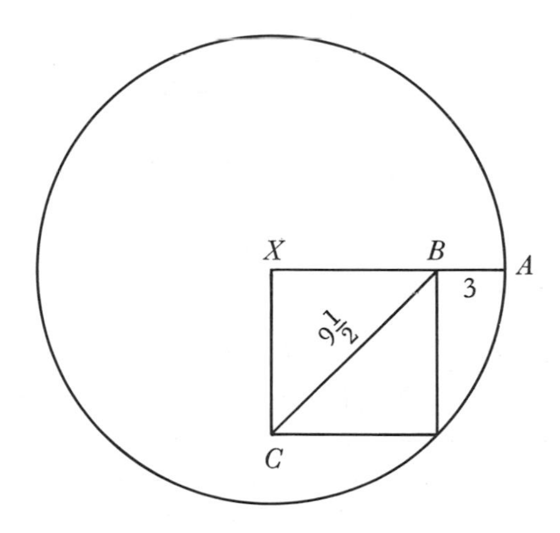

FIGURE 24

BA is 3 inches. What's the diameter of the circle—without using calculus, please?"

193° *Italian demonstration.* At a certain mathematics meeting, Professor Gian-Carlo Rota of M.I.T., in order to save scarce time and avoid lengthy detailed verbiage, carried out a demonstration with some waving of his arms. At the end he apologized for having to give an "Italian demonstration."

194° *Covering—open or closed?* A. N. Whitehead has said that it is the function of a teacher not to cover the subject, but to uncover it for the student.

At the conclusion of a mathematics course, a student approached the instructor, who shall be nameless, and said: "Professor X, I should like to comment on how well you covered the subject. Whatever you did not cover in class, you covered on the final examination."

ALAN WAYNE

195° *See?* A college instructor, asked to explain a problem, stood at the blackboard a moment, then wrote the answer. "I still don't get it," complained the student. The instructor erased the board, stood there longer, then carefully wrote the result. "No, I still don't get it," insisted the student. This next time the instructor wrote the statement of the problem, then stood facing it a long time. Finally, he wrote the answer again. The student shook his head. "But I've worked it for you three different ways!" shouted the instructor. (Compare with Item 360° of *In Mathematical Circles.*)—ALAN WAYNE

196° *A high school for mathematics.* Lewis Carroll, in a letter, once suggested some ideas that might be employed in the design of a school for the teaching of elementary mathematics. The letter may be found in W. F. White, *A Scrap-Book of Elementary Mathematics,* 4th edition, pages 201 and 202 (The Open Court Publishing Company, 1942). Subsequent additions and improvements to the proposed school were researched by CHED of Stratford-on-Penobscot. Some of the final results are offered here for whatever they may be worth.

The school building is to be composed of a basement and a first

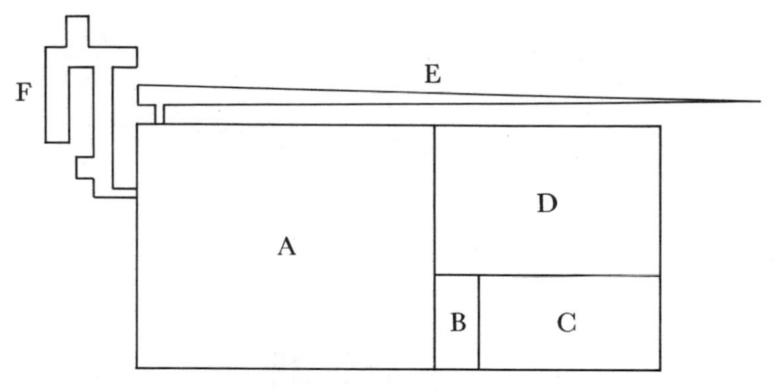

First Floor

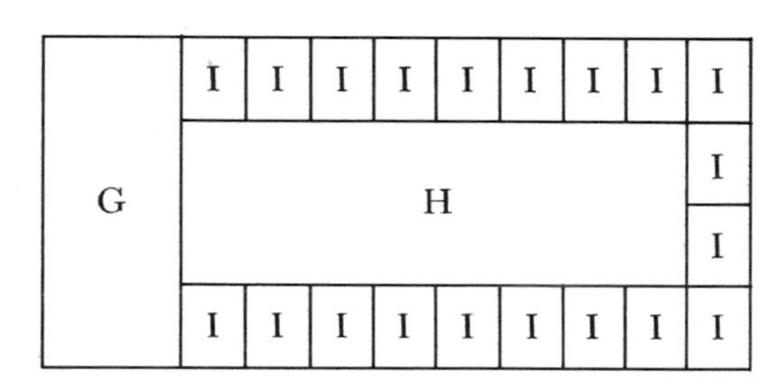

Basement

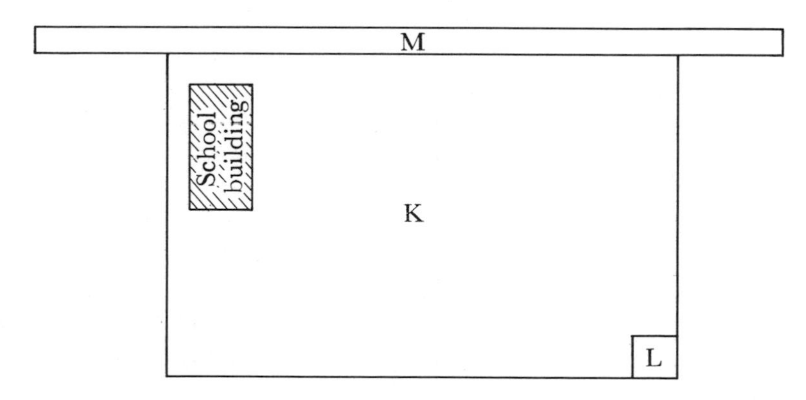

Athletic Fields

FIGURE 25

floor *only*, and is to be accompanied by appropriate athletic fields. Referring to Figure 25, the following conveniences are to be incorporated:

FIRST FLOOR

A. A very large room for calculating greatest common divisors.
B. A small nearby room where, with delicate tools, one may calculate least common multiples.
C A dark room fitted with a projector for showing repeating decimals in the act of repeating.
D. A room equipped with precision tools and balances for the purpose of dividing whole numbers.
E. An attached, long, and gradually tapering room for summing infinite geometric progressions, writing infinite decimal expansions, and finding limits of infinite sequences.
F. Another attached room of very irregular shape, for the study of inequalities.

BASEMENT

G. A padded and soundproof room where irrationals may be kept.
H. A large windowless room equipped with whips, chains, and racks for ruthlessly reducing fractions to lowest terms.
I. A number of small rooms off Room H, for storing the reduced fractions.

SECOND FLOOR

J. The second floor of this one-story school building is to be reserved exclusively for the study of imaginary numbers.

ATHLETIC FIELDS

K. A large piece of open ground where, in the afternoons with picks and shovels, the students can practice the extraction of roots.
L. A smaller, fenced-in area where the square roots may be kept so that their sharp corners will not injure the other roots.
M. A long narrow strip of ground, carefully leveled, for testing practically whether parallel lines do or do not meet; for this purpose the strip should reach "ever so far."

197° *Twenty-third slam.*

The college professor is my shepherd and I am in dire want;
He preventeth me from lying down in the bed which I renteth;
He leadeth me to distraction with his exam questions;
He shaketh my resolution to get a college degree;
He leadeth me to make a fool of myself before my classmates.
Yea, tho I burn my lamp until the landlady howleth, I fear much evil,
For he is against me.
His policies, his theories, and his ranting frighten my wits from me.
He assigneth me extra work in the presence of mine enemies;
He anointeth my quiz paper with red pencil marks,
And my zeros fill a whole column.
Surely, theories, exams, and themes will follow me all the days of my
college career,
And I will dwell in the bughouse forever.

ANONYMOUS

198° *Oral examination procedures.* Probably most of us in the college teaching profession have witnessed, at one time or another, some of the shameful and sadistic practices so prevalent in the oral examination procedures for Masters' and Doctors' degrees in mathematics. The examination often turns into a showoff performance by some member of the faculty and/or a medieval torture session at the expense of the unfortunate candidate. I recall one oral examination in which a faculty member flaunted his supposed brilliance for over an hour straight, and another examination whcrcin, after the miserable candidate was excused to wait outside the room while his examiners privately discussed his sad performance, cards were whipped out and a lengthy game played within the examination room.

Such procedures induced S. J. Mason several years ago to circulate the following rules and guidelines for conducting oral examinations in mathematics.

> The purposes of an oral examination are few and simple. In these brief notes the purposes are set forth and practical rules for conducting an oral examination are given. Careful attention to the elementary rules is necessary in order to assure a truly successful examination.

From the standpoint of each individual examiner the basic purposes of oral examination are:

(A) To make the examiner appear smarter and trickier than either (a) the examinee or (b) the other examiners, thereby preserving his self-esteem.

(B) To crush the examinee, thereby avoiding the messy and time-wasting problem of post-examination judgment and decision.

Both of these aims can be realized through diligent application of the following time-tested rules:

1. Before beginning the examination, make it clear to the examinee that his whole professional career may turn on his performance. Stress the importance and formality of the occasion. Put him in his proper place at the outset.

2. Throw out your *hardest* question *first*. (This is very important. If your first question is sufficiently difficult or involved, he will be too rattled to answer subsequent questions, no matter how simple they may be.)

3. Be reserved and stern in addressing the examinee. For contrast be very jolly with the other examiners. A very effective device is to make humorous comments to the other examiners about the examinee's performance, comments which tend to exclude him and set him apart (as though he were not present in the room).

4. Make him do it your way, especially if *your* way is esoteric. Constrain him. Impose many limitations and qualifications in each question. The idea is to *complicate* an otherwise simple problem.

5. Force him into a trivial error and then let him puzzle over it for as long as possible. *Just after* he sees his mistake but *just before* he has a chance to explain it, correct him yourself, disdainfully. This takes real perception and timing which can only be acquired with some practice.

6. When he finds himself deep in a hole, never lead him out. Instead, sigh and shift to a new subject.

7. Ask him side questions, such as, "Didn't you learn that in Freshman Calculus?"

8. Do not permit him to ask you clarifying questions. Never repeat or clarify your own statement of the problem. Tell him not to think out loud, what you want is the *answer*.

9. Every few minutes, ask him if he is nervous.

10. Station yourself and the other examiners so that the examinee cannot really face all of you at once. This enables you to bracket him with a sort of binaural crossfire. Wait until he turns away from you toward someone else, and then ask him a short direct question. With proper coordination among the examiners it is possible under favorable

conditions to spin the examinee through several complete revolutions. This has the same general effect as Item 2 above.

11. Wear dark glasses. Inscrutability is unnerving.

12. Terminate the examination by telling the examinee, "Don't call us; we will call you."

MATHEMATICIANS AND MATHEMATICS

199° *E. H. Moore's dictum on theory.* Professor E. H. Moore (1862–1932), in his lectures of 1906, said:

"We lay down a fundamental principle of generalization by abstraction:

"*The existence of analogies between central features of various theories implies the existence of a general theory which underlies the particular theories and unifies them with respect to those central features. . . .*"

200° *E. H. Moore's dictum on rigor.* Commenting on how rigorous a course in mathematics should be, Professor E. H. Moore said: "Sufficient unto the day is the rigor thereof."

201° *Rigor and rigor mortis.* It has been remarked that too much *rigor* in a mathematics course induces *rigor mortis* among the students.

202° *Gauss's second motto.* Gauss chose for his second motto* the following lines from *King Lear:*

> Thou, nature, art my goddess; to thy laws
> My services are bound. . . .

Gauss thus believed that mathematics, for inspiration, must touch the real world. As Wordsworth put it, "Wisdom oft is nearer when we stoop than when we soar."

203° *Modern mathematics and primitive ritual.* There is a striking similarity between modern mathematics and primitive ritual. In

---

* Gauss's first motto was: *Pauca sed matura* (few, but ripe). See Item 326° of *In Mathematical Circles.*

primitive ritual, objects representing certain things are treated with the feeling that the treatment is being performed on the things represented by the objects. For example, an image of an enemy may be stuck with pins, or dismembered, or smashed, in the belief that harm is being done to the enemy himself. In modern mathematics we have symbols representing things, and we operate on those symbols often with the feeling that we are operating on the things themselves.

204° *A mathematician's prayer.* If there is anything that might be called a mathematician's prayer, it must run something like this: "Oh Lord, please let me once more make a juicy discovery."

205° *A mathematician's brains.* Here is a story that made the rounds of the mathematics meetings a few years ago.

There was a surgeon who discovered how to remove a person's brains and to replace them with whatever kind of brains the patient desired. Of course, different kinds of brains cost different amounts of money.

One day a patient presented himself to the surgeon and said that he would like to change his brains. "Good," said the surgeon, "what kind of brains do you want? They come at various prices. For example, a lawyer's brains will cost you $10 an ounce, a judge's brains $50 an ounce, and so on."

"Oh, I don't want those kinds of brains," said the patient. "I would like to have the brains of a college professor."

"My, you do have expensive taste," replied the surgeon. "Now the brains of a college English professor will cost you $100,000 an ounce; those of a college history professor will cost you $200,000 an ounce. What kind of college professor's brains do you want?"

"I would like to have the brains of a college mathematics professor," asserted the patient.

"You do indeed have expensive taste," said the surgeon. "Those are the most expensive brains of all. They cost $1,000,000 an ounce."

"That is incredible," replied the patient. "Why should they cost so much? If a lawyer's brains are only $10 an ounce and a judge's brains only $50 an ounce, why should the brains of a college mathematics professor cost $1,000,000 an ounce?"

"Oh, I think you can see that," said the surgeon. "Just imagine how many mathematicians it takes to get an ounce of brains."

206° *Technique of problem solving.* In problem solving one must get more hose to get closer to the fire.

207° *Esthetic of problem solving.* In problem solving one should not drag out heavy artillery to take a small fort. Or, to change the metaphor, one should not use a sledge hammer to drive a thumbtack.

208° *Something to keep in mind.* Not only haven't all mathematicians been professors of mathematics, but all professors of mathematics haven't been mathematicians.

209° *An impartial account of Western mathematics.* Individual envy, national jealousy, and racial fanaticism have often marred the history of mathematical discovery and invention. Commenting on this point, E. T. Bell once wrote: "An impartial account of Western mathematics, including the award to each man and to each nation of its just share in the intricate development, could be written only by a Chinese historian. He alone would have the patience and the detached cynicism necessary for disentangling the curiously perverted pattern to discover whatever truth may be concealed in our variegated occidental boasting."

210° *An historic moment in mathematics.* Here is another charming quote from Alfred North Whitehead's *An Introduction to Mathematics* (1911): "It is impossible not to feel stirred at the thought of the emotions of men at certain historic moments of adventure and discovery—Columbus when he first saw the Western shore, Pizarro when he stared at the Pacific Ocean, Franklin when the electric spark came from the string of his kite, Galileo when he first turned his telescope to the heavens. Such moments are also granted to students in the abstract regions of thought, and high among them must be placed the morning when Descartes lay in bed and invented the method of coordinate geometry."

211° *A principle of discovery.* There is a remarkable principle of discovery and advancement in mathematics and science, namely, the *constructive* doubting of some traditional belief. When Einstein was asked how he came to invent the theory of relativity he replied, "By challenging an axiom." Lobachevsky and Bolyai challenged Euclid's axiom of parallels; Hamilton and Cayley challenged the axiom that multiplication is commutative; Lukasiewicz and Post challenged Aristotle's axiom of the excluded middle. Similarly, in the field of science, Copernicus challenged the axiom that the earth is the center of the solar system; Galileo challenged the axiom that the heavier body falls the faster; Einstein challenged the axiom that of two distinct instants one must precede the other. This constructive challenging of axioms has become one of the commoner ways of making advances in mathematics, and it undoubtedly lies at the heart of Georg Cantor's famous aphorism: "The essence of mathematics lies in its freedom."

Of course, the doubt or the challenge must be a *constructive* one. That is, the doubted or challenged belief must not merely be discarded, but must be replaced by an alternative belief, and then the consequences of this alternative worked out.

212° *Mathematical density.* Since 1960, Hungary has been reported to have more mathematicians per capita than any other country of the world.

213° *D'Alembert on mathematics.* Perhaps the most perceptive of d'Alembert's comments on mathematics is the following: "I have no doubt that if men lived separate from each other, and could in such a situation occupy themselves about anything but self-preservation, they would prefer the study of the exact sciences to the cultivation of the agreeable arts. It is chiefly on account of others that a man aims at excellence in the latter; it is on his own account that he devotes himself to the former. In a desert island, accordingly, I should think that a poet could scarcely be vain, whereas a mathematician might still enjoy the pride of discovery." (In connection with mathematics and desert islands, also see Item 128°.)

214° *Dugald Stewart on mathematics.* The Scottish philosopher

Dugald Stewart (1753–1828) was the son of Matthew Stewart, professor of mathematics at the University of Edinburgh. In 1772, when ill health forced his father to stop teaching. Dugald took over the mathematics classes at Edinburgh, and three years later he became joint professor with his father. In 1785 he was appointed professor of moral philosophy. Dugald was a gifted orator and attained great popularity as a lecturer. In his *Elements of the Philosophy of the Human Mind* (1792–1827), he makes the following discerning remark about mathematics, which reminds us of the somewhat similar comment made earlier by d'Alembert and reported here in the preceding Item.

"Whoever limits his exertions to the gratification of others, whether by personal exhibition, as in the case of the actor and of the mimic, or by those kinds of literary composition which are calculated for no end but to please or to entertain, renders himself, in some measure, dependent on their caprices and humours. The diversity among men, in their judgments concerning the objects of taste, is incomparably greater than in their speculative conclusion; and accordingly, a mathematician will publish to the world a geometrical demonstration, or a philospher, a process of abstract reasoning, with a confidence very different from what a poet would feel, in communicating one of his productions even to a friend."

215° *Thomas Hill on mathematics.* This seems a proper place for a particular quote from Thomas Hill (1818–1891) on mathematics. Hill was at one time president of Harvard College; he is known in mathematics chiefly for his work on intrinsic coordinates.

"There is something sublime in the secrecy in which the really great deeds of the mathematician are done. No popular applause follows the act; neither contemporary nor succeeding generations of the people understand it. The geometer must be tried by his peers, and those who truly deserve the title of geometer or analyst have usually been unable to find so many as twelve living peers to form a jury. Archimedes so far outstripped his competitors in the race, that more than a thousand years elapsed before any man appeared, able to sit in judgment on his work, and to say how far he had really gone. And in judging of those men whose names are worthy of being mentioned in connection with his—Galileo, Descartes, Leibniz, Newton, and the mathematicians created by Leibniz and Newton's calculus—we are forced to depend upon their testimony of one another. They are too far above our reach for us to judge them."

216° *The mathematician and his methods.* Ludwig Boltzmann (1844–1906) once commented that analytic geometry seems almost cleverer than the man who invented it. This is a feeling one experiences in connection with practically every highly ingenious mathematical method. For an allied observation, concerning a formalist and his pencil, see Item 251° of *In Mathematical Circles.*

217° *The researchers and the teacher.* Several creative mathematicians and a teacher of mathematics were eating lunch together when one of the creative men queried: "Suppose that, in imitation of Archimedes, each one of us should request some indication of his proudest professional achievement be engraved on his tombstone, what would each of us choose?" One researcher said he would pick a certain remarkable formula in analysis that he had discovered, another chose the basic figure in one of his geometrical investigations, still another wanted a brief statement of one of the fine theorems he had managed to prove, and so on, until each researcher had expressed his wish. Then, with a smile of amusement, one of the researchers turned to the teacher and asked: "And what would you ask to have engraved on your tombstone?" As the group of researchers settled back to enjoy the ensuing "fun," the teacher replied: "I would ask that the names of my five best students be engraved on my tombstone; that would represent my greatest achievement." Somehow, the anticipated "fun" did not materialize.

218° *Mental mathematics.* In the Institute for Advanced Study, at Princeton, New Jersey, two research mathematicians sat staring at a blackboard, each with a piece of chalk in his hand. The blackboard was blank. After an hour's complete silence, one of them looked up and nodded his head vigorously, in triumph. The other looked at him, then shook his head, sadly, in contradiction.—ALAN WAYNE

219° *Test question.* The guidance counselor in a university uses the following question to help him advise students as to the course of study they should take up. "A cow has four legs. If the tail is counted as a leg, how many legs would the cow have?"

The student who answers, "Four. Calling a tail a leg does not make it one," is a realist, and is guided into science.

The student who answers, "Five," shows imagination, and is guided into mathematics.

The student who answers, "Three," shows ability in juggling facts, and is guided into social studies.

The student who answers, "That is a good question," is guided into teaching.—ALAN WAYNE

220° *Occupational test.* A big corporation classifies its job applicants into engineers and mathematicians on the basis of a two-part test.

On Part I of the test the candidate is shown a chair, a table, and a stove with a lit burner. On the table is a kettle full of water. The problem is to boil the water. If the candidate takes the kettle and puts it on the lit burner, he has passed Part I of the test.

On Part II of the test, the candidate is shown the same set-up, except that now the kettle is on the chair, instead of on the table. The problem again is to boil the water. The candidate who takes the kettle and puts it on the lit burner is immediately classed as an engineer. The candidate who takes the kettle from the chair and puts it on the table, saying, "Now it's the same problem as the previous one, which I've already solved!"—he's the mathematician!—ALAN WAYNE

221° *The mathematician.* Somewhat typical is the mathematician presenting a learned paper before a meeting of the American Mathematical Society. At one point of his presentation he is thinking of the number "five." But he says "four" and writes "three" on the blackboard. The correct number, of course, is "six."

222° *A place for credulity.* Morris Kline has commented on how fortunate it was that the mathematicians of the seventeenth century "were so credulous and even naive, rather than logically scrupulous. For the greatest period of mathematical creativity was already under way and free creation must precede formalization and logical foundations."

223° *Five quotes from Novalis.* Friedrich Leopold Hardenberg (1772–1801), better known by his pen name Novalis, was a German author and hymnologist and, with the Schlegels and Tieck, one of the founders of the German Romantic School. His hymns and fragmentary novels are characteristic products of the School. In addition to philosophy, art, religion, and law, he was thoroughly grounded in mathematics and the natural sciences, and a surprising number of shrewd comments bearing on mathematics and science can be found in his works. It might well be that one could assemble a little booklet of such penetrating and discerning remarks. Here are a few brief ones concerning mathematics:

1. The real mathematician is an enthusiast *per se*; without enthusiasm, no mathematics.
2. One may be a mathematician of the first rank without being able to compute; it is possible to be a great computer without having the slightest idea of mathematics.
3. Music has much resemblance to algebra.
4. What logarithms are to mathematics, that mathematics are to the other sciences.
5. Pure mathematics is the real magician's-wand.

224° *Teaching versus research.* "Christ was a great teacher," someone remarked. "Yes," replied another, "but what did he publish?"

225° *Publication and reward.*

A theorem a day
Means promotion and pay!
A theorem a year
And you're out on your ear!

PAUL ERDÖS

226° *A hazardous occupation.* The following item appeared in the *Evening Express* of Portland, Maine, February 12, 1970.

PHILADELPHIA (UPI)—A mathematician who apparently felt his University of Pennsylvania professors were denying him his doctorate shot his adviser and his former department head during a lecture Wednesday, then fired a bullet into his own mouth.

Police said Robert H. Cantor, 33, stood in the doorway of the lecture-hall and fired five shots from a .45-caliber automatic at Dr. Walter Koppelman, 40, and Dr. Oscar Goldman, 45.

Without saying a word, Cantor turned and placed the weapon into his mouth, pulled the trigger and fell dead in a pool of blood in a hallway five feet from the doorway of the lecture hall in the David Rittenhouse Laboratory, the mathematics building.

Goldman, seated in the front row of a six-row tier of seats, fell to the floor with bullet wounds of the right wrist and foot and left hand.

Goldman was later listed in satisfactory condition at the University of Pennsylvania Presbyterian Hospital.

Koppelman, sitting two seats away from Goldman, was struck in the right arm. Doctors said the bullet entered his chest. He was hospitalized in serious condition before lengthy surgery.

Philadelphia Police Sgt. Kenneth Schwartz said Cantor apparently harbored a grudge against several professors in the Penn Mathematics Department because he has been unable to earn his doctorate.

"He held a grudge against the faculty," Schwartz said. "He had to submit theses for his doctorate. He'd keep submitting them and they'd keep turning them down."

> Koppelman was Cantor's thesis adviser. Goldman was a professor and a former chairman of the Mathematics Department from 1963 to 1967.
>
> A Temple professor attending the lecture, Dr. Emil Grosswald, said he knew Cantor when he taught him at Penn as "a very nice boy, but at times he was erratic."

The final outcome of the affair was that Dr. Goldman pulled through, but Dr. Koppelman died about a month later. Had Dr. Koppelman lived he would have been paralyzed from the waist down.

WOMEN OF MATHEMATICS

It is only quite recently that women have seriously entered the fields of mathematics and the exact sciences. Not long ago such pursuits by women were frowned upon, and opportunities for such pursuits by women were virtually nonexistent. This state of affairs is reflected in the history of mathematics, so that from the closing days of Greek antiquity to the early twentieth century it is possible to list only five truly outstanding women mathematicians: Hypatia of Alexandria (*ca.* 370–415), Maria Gaetana Agnesi (1718–1799), Sophie Germain (1776–1831), Sonja Kovalevsky (1850–1891), and Emmy Noether (1882–1935). It is expected that present-day female emancipation and recognition of the basic equality of the sexes will produce a growing number of distinguished women mathematicians.

We here give a few stories about Sophie Germain, Sonja Kovalevsky, and Emmy Noether; Hypatia and Maria Agnesi have already been considered in Items 90°, 273° and 274° of *In Mathematical Circles*. Of the nine stories selected, the first eight are adapted, with permission, from the splendid article by Professor Rora F. Iacobacci entitled "Women of mathematics," which appeared as the special feature in the April, 1970, issues of *The Mathematics Teacher* and *The Arithmetic Teacher*.

227° *How Sophie Germain was led to the study of mathematics.* Sophie Germain was born in Paris on April 1, 1776. As a child she had not given any evidence of an extraordinary vocation, but in 1789, when she was barely thirteen, a fortuitous coincidence of events led her to the serious study of mathematics.

The meeting of the Estates General in May 1789 had brought to one focal point all the economic, social, and political conflicts of eighteenth-century France. In June the Third Estate declared itself the National Assembly, and on July 14th the Bastille fell; a decade of revolutionary violence lay ahead. Confined to her house, Sophie would spend long hours reading in her father's library. One day she came upon the legend of the death of Archimedes in Montucla's *History of Mathematics*. Here she read how Archimedes, a native of the Greek city-state of Syracuse on the island of Sicily, aided in the defense of that city against the Romans by devising machines to repel the enemy, and how, when the city finally fell in 212 B.C., he was speared down by a Roman soldier while engrossed in the study of a mathematical figure in the sand. This story impressed Sophie deeply. Here was a science, mathematics, that was capable of absorbing the spirit so completely as to make it totally oblivious to all its surroundings. Such a strong and sustaining study would aid her greatly in the face of all the torment present in her city. She resolved to study mathematics.

RORA IACOBACCI

228° *Sophie Germain as M. Le Blanc.* When the École Polytechnique opened in 1794 it did not accept women, but Sophie was able to procure the lecture notes of various professors. The new analysis of Lagrange naturally engaged her attention, and taking advantage of the opportunity provided to students of submitting written observations to the professor at the end of the course, she communicated hers to Lagrange under the pseudonym of M. Le Blanc. Lagrange praised the comments, and on learning the identity of the author, commended the young analyst. From that time on, Sophie saw herself as a mathematician.

In 1801 Gauss's *Disquisitiones arithmeticae* was published, and in 1804 Sophie entered into correspondence with Gauss by sending him some of the results of her arithmetical investigations, again under the pseudonym of M. Le Blanc. After several letters, Gauss became impressed by the sagacity of many of her observations, and when her identity was revealed to him by the French general Pernety, to whom Sophie had commended Gauss, he did not hesitate to compliment her.

RORA IACOBACCI

229° *How Sonja Kovalevsky was led to the study of mathematics.* Sophia Korvin-Krukovsky, later known by the name Sonja Kovalevsky, was born in Moscow on January 15, 1850, into a family of the Russian nobility. When she was eight years old the family moved to the country estate at Polibino, and it was there that she spent her childhood and received her early education. Her first teacher on general subjects described her as a sweet, attractive child of rather sturdy build, with hazel-brown eyes that sparkled with intelligence and kindliness. Her mathematical studies consisted of arithmetic for two and a half years, algebra and geometry for three and a half years, and finally plane and solid geometry.

Sonja has written of two factors that attracted her to the study of mathematics. The first was her Uncle Pyotr, who had studied the subject on his own and would speak of squaring the circle and of the asymptote (that line that approached the curve nearer and nearer but never met it), as well as of many other things that excited her imagination. The second was a curious "wallpaper" that was used to cover one of the children's rooms at Polibino, which turned out to be lecture notes on differential and integral calculus that had been purchased by her father in student days. These sheets fascinated her and she would spend hours trying to decipher separate phrases and to find the proper ordering of the pages.—RORA IACOBACCI

230° *How Sonja Kovalevsky got to study in Germany.* In the autumn of 1867 Sonja went to St. Petersburg where she studied calculus with Alexander Strannolyubsky, a teacher of mathematics at the naval school. While there, she consulted the prominent Russian mathematician Chebyshev about her mathematical studies, but since Russian universities were closed to women, there seemed to be no way that she could pursue advanced studies in her native land.

The situation confronting Sonja at this time was also plaguing hundreds of other girls of the best Russian families who wished to continue their studies and to devote their best powers to the progress of their country. Since the only solution was to study at a foreign university, and since this idea usually met with parental opposition, many girls contracted nominal marriages with young men who shared their views. The "wife" would then be free to study abroad. Sonja's sister Anna knew a person who was willing to assist them. He was Vladimir Kovalevsky, who later became a distinguished paleontologist.

Vladimir was very impressed with Sonja's advanced mathematical studies, her fluency in languages, and her overall talents and industry. They were married in September 1868, and went to Heidelberg the next spring.—Rora Iacobacci

231° *Weierstrass's favorite pupil.* At Heidelberg (1869–1870) Sonja Kovalevsky heard Konigsberger and Du Bois-Reymond lecture on mathematics and Kirchoff and Helmholtz lecture on physics. However, Konigsberger was a former student of Karl Weierstrass (1815–1897) and, having caught her teacher's enthusiasm for the master, Sonja wished to attend the lectures of the famous mathematics professor at Berlin. She arrived there for the new term in August 1870, but found that the University, which did not accept women, would make no exception for her. She then approached Weierstrass directly, and he, upon receiving a reassuring recommendation of Sonja from Konigsberger, offered to give her private lessons.

Sonja quickly became Weierstrass's favorite pupil. He repeated his university lectures to her, and discussed his unpublished works, the latest scientific achievements, problems of stability, and new theories of geometry with her. He wrote a friend that he had had very few pupils who could be compared with her for industry, ability, diligence, and passion for science. She studied with him for four years, 1870–1874, and during that time not only covered the university course of mathematics, but wrote three important works: (1) On the Theory of Partial Differential Equations; (2) On the Reduction of Abelian Integrals of the Third Kind to Elliptic Integrals; and (3) Supplementary Research and Observations on Laplace's Research on the Form of the Saturn Ring.

In 1874, Göttingen University conferred on Sonja the degree of Doctor of Philosophy, *in absentia*, excusing her from the oral examination because of the remarkable excellence of the three papers she had sent in. She had submitted the paper on partial differential equations as a thesis, and presented the other two in addition.—RORA IACOBACCI

232° *Sonja Kovalevsky's motto.* In 1888, at the age of thirty-eight, Sonja achieved her greatest success when she was awarded the *Prix Bordin* by the Institut de France for her memoir *On the Problem of the Rotation of a Solid Body about a Fixed Point.* The paper had been submitted with the motto: "Say what you know, do what you must, come what may." It was not only judged the best of the fifteen papers submitted, but was considered to be of such exceptional merit that the prize money was raised from 3000 to 5000 francs.—RORA IACOBACCI

233° *Why Emmy Noether became an algebraist.* The two strongest mathematical influences on Amalie Emmy Noether (1882–1935) in her early life were her father Max Noether (1844–1921), a distinguished mathematician at the University of Erlangen, who played an important role in the development of the theory of algebraic functions, and the mathematician Paul Gordon (1837–1912), who was also associated with the University and became a close family friend. A bit of diversity enters here: although both men were specialists in algebra, they were essentially different kinds of mathematicians. Max Noether was strong on structure, while Gordon was an "algorithmiker." Emmy wrote her doctoral thesis under Gordon in 1907, "On Complete Systems of Invariants for Ternary Biquadratic Forms," and it is entirely in line with the Gordon spirit. As an extreme example of formal computations, it contrasts sharply with her works in maturity, which are extreme examples of conceptual axiomatic thinking in mathematics.

When Gordon retired in 1910, he was followed in the second year by Ernst Fischer, whose field was again algebra and, in particular, the theory of elimination and of invariants. His influence upon Emmy was quite penetrating, and under his direction the transition from the formal standpoint to the Hilbert method of approach was accomplished. This period extended to about 1919.—RORA IACOBACCI

234° *Einstein's tribute to Emmy Noether.* In the spring of 1933 Germany was in the grips of a national revolution, and Emmy Noether, as well as many others, was prohibited from participation in all academic activities. The impossibility of working at Göttingen brought her to Bryn Mawr College in Pennsylvania as a professor, and to the Institute for Advanced Study in Princeton. In America, Emmy once again found respect and friendship among her peers, and appreciation and enthusiasm among her students, all of which contributed towards making her last year and a half both very happy and productive. She died on April 14, 1935, at the age of fifty-three and at the height of her career. In his tribute to Emmy Noether in May 1935, Albert Einstein wrote:

> In the judgment of the most competent living mathematicians, Fräulein Noether was the most significant creative mathematical genius thus far produced since the higher education of women began. In the realm of algebra in which the most gifted of mathematicians have been busy for centuries, she discovered methods which have proved of enormous importance in the development of the present day younger generation of mathematicians.

RORA IACOBACCI

235° *Landau on Emmy Noether.* Someone once described Emmy Noether as the daughter of the German mathematician Max Noether. To this Edmund Landau replied: "Max Noether was the father of Emmy Noether. Emmy is the origin of coordinates in the Noether family."

WHEREIN THE AUTHOR IS INVOLVED

OVER a long period of teaching mathematics, editing sections of mathematics journals, delivering addresses around the country, collecting mathematical items, and corresponding with colleagues, some curious or amusing, but apropos, incidents are bound to occur. They did, and here are a few of them.

236° *A pair of old debts.* During an intermission of the NCTM meeting held in Corpus Christi a couple of years ago, I found myself at a little table enjoying a Coke with Father Bezuska. The conversation turned to geometry and the old master, Euclid. I finally said, "I owe an immeasurable debt to Euclid. Reading the first six books of his

Elements in school marked an important turning point in my life, for it decided that I would go into mathematics as my life's work." And I concluded by musing, "Isn't it remarkable that what a man did some 2000 years ago should so effect one's life?" And then I suddenly realized that perhaps there was nothing so remarkable about it after all, for there sitting across the table from me was Father Bezuska in clerical dress, and he similarly had had his life markedly effected by the doings of a man who lived about 2000 years ago.

237° *In memory of Dr. William D. Taylor.* One of my earliest teaching appointments was that of Assistant Professor of Applied Mathematics at Syracuse University. In addition to teaching various courses in engineering mathematics, I was asked to teach a course in mechanics. Since I had had little experience with mechanics, I decided to help prepare myself by sitting in on the elegant lectures on the subject given at Syracuse by the lovable Professor William E. Taylor (1870–1945). I shall never forget the opening class meeting. Professor Taylor, who was a deeply religious man, commenced the course with a solemn prayer in which he asked the Lord to help his students master the beautiful subject of mechanics, so that they might emerge from the course with good grades and an abiding appreciation of the material. Though it turned out that the first wish was not granted to every student, surely the second one was.

238° *The mathematician and the fundamentalist.* At one time I lived directly across the street from a church officiated by a fundamentalist minister who became concerned about my life in the hereafter. He told me grim stories about the lower regions and described the torrid horrors I might expect if I didn't snap to and change my ways by joining his church. One regretful day, in a jest that he completely failed to appreciate, I told him that I had no fear of the lower regions, because all the mathematicians and engineers who have gone there have undoubtedly remarkably improved the place with air conditioning and refrigeration.

Recently I received the following item clipped from the Bangor *Daily News*, Thursday, January 8, 1970.

HAD TO HAPPEN

HELL, Norway (UPI)—The water froze in Hell Wednesday when the temperature dropped to 6 degrees below zero.

Too many engineers, I suppose, fooling around with the air conditioning. I do wish they would be more careful; they might spoil a good thing.

239° *A Hoggatt witticism.* The lovable, energetic, and talented west-coast mathematician V. E. Hoggatt, Jr., throws out amusing witticisms with seeming ease. Just after I informed him I was in the process of making this second trip around the "mathematical circle," I received the following from him by air mail:

"Do you think that persons who are going in circles should be set straight? Yes? Then wouldn't they be off on a tangent?"

I have been wondering exactly what he was trying to tell me.

240° *All good books are read.* One of my greatest teaching pleasures was meeting Verner Hoggatt when he was a student in college. He was in my freshman class at the University of Puget Sound and then later took a number of courses from me at Oregon State University. I've never had a more enthusiastic student, nor one I enjoyed so much. It was the start of a long, warm, and valued friendship. Verner Hoggatt, now a Professor of Mathematics at San Jose State College, is today recognized as probably the world's foremost authority on the Fibonacci and allied sequences, and is the energetic editor-in-chief of the remarkable *Fibonacci Quarterly*, a journal which he founded.

One of Verner's many charms is a ready and delightful wit. I recall an early instance, dating from that freshman class at Puget Sound. I was teaching the principle of mathematical induction and I said to the class: "Imagine a row of books on a bookshelf, and suppose we somehow know that if one book is red, then the book next following it is also red. Peeking through a crack we observe that the eighth book is red. What can we conclude?" At this Verner asked, "Are they all good books?" "All right," I said, "let us assume they are all good books." "Then," he replied, "*all* the books on the shelf are red." A bit shocked, I asked, "Why is that?" "Because, you see *all* good books are read," he explained.

241° *Moving to Gorham.* Arrangements were made for me to spend a year or two assisting the Mathematics Department of Gorham State College, one of the units of the newly formed all-state University of Maine. Soon after the completion of the arrangements, a short two-column item, released by the public relations office of the University of Maine, appeared in the Bangor *Daily News.* It contained a picture of me with a brief statement that I and my family were temporarily moving from Orono to Gorham. This little item was boxed on the top and right side by a much longer, four-column communication bearing a large headline that ran across all four columns and that boldly asserted: NEW HOME FOUND FOR THE INSANE.

242° *Adam and Eve.* There is a venerable and much admired Dr. Adams in Orono who for years has administered to the medical needs of the community. One evening my home telephone rang and I answered it. A woman's voice inquired, "Is this Dr. Adams?" "No," I replied, "this is Dr. Eves." There followed a long silence and finally the lady's voice returned, saying, in exasperation, "Oh, come now!" On this she abruptly hung up, leaving me quite perplexed for several moments.

243° *Mr. S. T. Thompson.* Many amusing things occur over a long tenure of editorship, at least so it happened to me during my more than twenty-five years as editor of the Elementary Problem Department of *The American Mathematical Monthly.* An interesting instance is the case of Mr. S. T. Thompson of Tacoma, Washington.

The Thompson case came about in this way. On occasion I found in my editorial work that I was able to discover a considerably better solution to a problem than any of those that were submitted. Indeed, sometimes my solution was the only one I possessed. Feeling it somewhat unethical to publish my own solutions, I invented a mythical Mr. S. T. Thompson, placed him vaguely in Tacoma, Washington (where I was living at the time), and gave him credit for my solutions.

Now Professor Harry Gehman, who for many years was the energetic and superb Secretary-Treasurer of the Mathematical Association of America, was ever alert for possible new members of the Association. Seeing Mr. S. T. Thompson's name appearing now and then among the solvers of the *Monthly's* problems, Harry wrote to me asking for Thompson's complete address, as he wished to invite Thompson to become a member of the Association. I wrote back to Harry that there was no use in inviting Mr. Thompson to become a member, for Mr. Thompson had an insuperable dislike of the Association and absolutely nothing would ever induce him to join. Needless to say, Harry found this attitude puzzling, frustrating, and quite incomprehensible.

This is the first time the true state of affairs concerning Mr. S. T. Thompson has been publicly divulged.

244° *Professor Euclide Paracelso Bombasto Umbugio.* Another case, somewhat similar to that of Mr. S. T. Thompson, originated in the pages of the Elementary Problem Department of *The American Mathematical Monthly* back in 1946. At that time, Professor George Pólya and I thought it might be enlivening if in each April issue of the journal there appeared a sort of April fools problem—a problem for which a straightforward solution is tedious and long, but for which with cleverness one can find an extremely brief and elegant solution. It was decided that these problems would emanate from a windy, verbose, but kindly numerologist, Professor Euclide Paracelso Bombasto Umbugio of Guayazuela.

Professor Umbugio's fame spread and in time April fool proposals were submitted through him by good bona fide mathematicians. The hoax fooled many readers of the *Monthly*, and the letters received asking for his address (so that some scientific correspondence could ensue) would fill a little pamphlet. Professor Leo Moser, in his graduate-school days, was among those seeking Professor Umbugio's address. All inquirers were informed that Professor Umbugio moved about so much that the best way to reach him was through correspondence sent in care of the editor of the Elementary Problem Department.

245° *An amusing incident in the life of an editor.* There is an elusive and tantalizing little problem in elementary geometry known as the "butterfly problem": *Let O be the midpoint of a given chord of a circle, let two other chords TU and VW be drawn through O, and let TW and VU cut the given chord in E and F respectively; prove that O is the midpoint of FE.* The problem receives its name from the fancied resemblance of the figure of the problem (see Figure 26) to a butterfly with outstretched wings, and

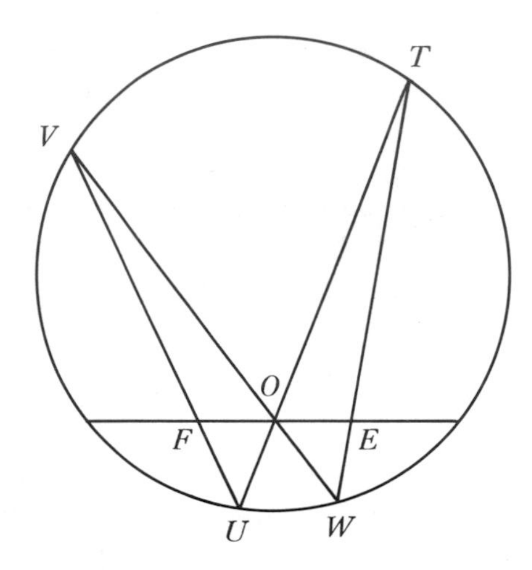

FIGURE 26

it is a very unusual student of high school geometry who succeeds in solving the problem.

In an effort to obtain a number of different but elegant solutions to the butterfly problem, I once ran it in the Elementary Problem section of *The American Mathematical Monthly* (Problem E 571, May 1943). Some truly ingenious solutions were submitted, and in due time (the February 1944 issue) I published a number of them. Among the published solutions was a particularly attractive one by Professor E. P. Starke, one of our country's outstanding problemists and the editor, at the time, of the Advanced Problem section of the *Monthly*.

Then a few years passed by, and one day I received a letter from Professor Starke stating that that semester he had a bright Norwegian student in one of his classes who gave him a very troublesome geometry problem. Professor Starke confessed that he had spent an inordinate amount of time on the problem, but without success, and he wondered if I could perhaps help him out with it. He then stated the problem, and lo, it was the butterfly problem! For a solution I referred him, without mentioning any names, to those pages in *The American Mathematical Monthly* wherein his own elegant treatment had appeared.

246° *Securing Wiener's hat for my mathematical museum.* One time a small group of graduate students accompanied some of the professors of Harvard and M.I.T. to New Haven to attend a mathematics meeting at Yale University. I was among the graduate students and Norbert Wiener was among the professors. The day was rainy and we all wore hats to keep the rain from running down our necks. About noontime our group walked over to a cafeteria at Yale for lunch and we hung our raincoats and hats on a clothes tree in the front part of the cafeteria. After lunching, we rose to don our coats and hats. Professor Wiener unthinkingly picked my hat (which in no way resembled his own) and placed it on his head. Since my hat was considerably smaller than his, I felt sure he would soon notice his error, but he didn't, and he stood there with my small hat ridiculously balanced on the top of his head. I accordingly selected his large hat and put it on my head, it coming down over my ears and resting on the bridge of my nose. Thus capped I looked up at Professor Wiener, feeling certain that now he would see that something was wrong. He noticed nothing, not even the loud mirth of the others of our group, and thus attired we marched out of the cafeteria with traded hats.

247° *Securing a lock of Einstein's hair for my mathematical museum.* I had the very good fortune to spend a year of graduate study at Princeton University. It was during that year that Dr. Albert Einstein became a member of the Institute for Advanced Study at Princeton, and I got to know the great man. Since I lived at the top of a short tower in the graduate school, and Einstein lived in the village a couple of blocks beyond, I occasionally walked Einstein to or from the university. On one rainy day I fell in step with Einstein on the way to the university, and during the walk the pelting rain battened some of his considerable mane (he very seldom wore a hat) down over his forehead. We proceeded to Einstein's office in Fine Hall, discussing something, and he let his hair dry in its annoying position. Bothered by the hair, he would every now and then swipe at it with his hand in an effort to brush it up, but it only fell back over his forehead. Suddenly, in a little petulant fit, he rushed over to a desk, secured a pair of scissors, cut off the offending lock and threw it in the waste basket. Always thinking about my mathematical museum, I later extracted the hair from the basket and reverently put it in an envelope.

248° *Securing Hardy's scarf for my mathematical museum.* During my year of graduate study at Princeton University, Professor Hardy came over from England to spend a semester lecturing on some Tauberian theorems. He soon had us all charmed.

Hardy wore heavy tweed suits, and even in the coldest weather augmented his dress with only a very long dirty white scarf that he wrapped round and round his neck. It wasn't long before I coveted that scarf for my mathematical museum. So, one day, after Hardy had gone into the great Gothic dining hall where we ate suppers, I took his long scarf off the hook where he had hung it and replaced it with the nicest and most expensive scarf I had been able to purchase in the village. After supper, when Hardy came out to reclaim his scarf, he found the beautiful one in its place. He began grumbling, but I soothed him by pointing out that whoever had taken his scarf had certainly left a considerably nicer one in its place. He conceded this, and finally took the new scarf in place of the old one. I waited several days to be sure he was satisfied with the surreptitious exchange. Finding him quite content, I entered his scarf as a new acquisition into my mathematical museum—in its original dirty and unkempt condition, of course.

249° *A doubtful acquisition for my mathematical museum.* It was in the summer of 1960 that the great American geometer Oswald Veblen passed away at his summer cottage near Brooklyn, Maine. The following summer I visited the cottage, much as one would visit a sacred shrine, and I there met Mrs. Veblen, who was about to take some lady friends on a little automobile excursion. She graciously gave me permission to wander about the grounds while she was gone, and even left the cottage unlocked in case I wished to explore inside. I declined the invitation to go inside, but I did walk around the rather extensive grounds, from the cottage, which was set well in from the road, down to the seashore, where the view of distant Acadia National Park was superb. In my wandering about, I came across a full-length bright-yellow lead pencil, broken in the middle but still adhering together, as though it had been stepped upon and lain on the ground all winter. I picked the pencil up, musing to myself as to whether it might not have belonged to and been used by the master himself. For, I reasoned, a geometer almost always carries a pencil with him, and if Mrs. Veblen were to use a pencil she would in all likelihood use it indoors to write out a grocery order or some such thing. At length I pocketed the pencil as a possible acquisition for my mathematical museum.

The next summer I happened to teach in an NSF Summer Institute at Bowdoin College, and Professor Tucker of Princeton was also there. During a luncheon together I told Professor Tucker the story of the broken pencil. He immediately settled into a concentrated study, apparently trying to figure out how he might discredit the acquisition. Finally, brightening up, he asked me a startling question. "Did you know," he triumphantly asked, "that during his last years Professor Veblen was almost totally blind—that he possessed only a peripheral vision?" I hadn't known this, and for a moment it looked as though the pencil would have to be discarded. But then it occurred to me: who but a person with only peripheral vision would drop a full-length bright yellow pencil on the ground and then not be able to find it again? When I disclosed this line of reasoning to Professor Tucker, he subsided into a seemingly disappointed silence.

NICOLAS BOURBAKI

SINCE 1939 a comprehensive set of volumes on mathematics, starting with the most general basic principles and proceeding into various specialized areas, has been appearing in France under the alleged authorship of a Nicolas Bourbaki. Bourbaki first appeared in connection with some notes, reviews, and other papers published in the *Comptes Rendus* of the French Academy of Sciences, and elsewhere. Then began the piecemeal construction of Bourbaki's major treatise. The purpose of the major treatise was explained in a paper that was translated into English and published in 1950 in *The American Mathematical Monthly* under the title, "The architecture of mathematics."

A footnote to this paper reads: "Professor N. Bourbaki, formerly of the Royal Poldavian Academy, now residing in Nancy, France, is the author of a comprehensive treatise of modern mathematics, in course of publication under the title *Éléments de Mathématique* (Hermann et Cie, Paris, 1939–), of which 10 volumes have appeared so far." More than thirty volumes have now (1970) appeared.

250° *A polycephalic mathematician.* Nicolas Bourbaki's name is Greek, his nationality is French, and he must be ranked as one of the most influential mathematicians of our century. His works are much read and quoted. He has enthusiastic supporters and scathing critics. And, most curious of all, he does not exist!

For Nicolas Bourbaki is a collective pseudonym employed by an informal group of mathematicians. Though the members of the organization have taken no oath of secrecy, it has amused most of them to be somewhat cryptic about themselves. Nevertheless, their names are largely an open secret to most mathematicians. It is believed that among the original members were C. Chevalley, J. Delsarte, J. Dieudonné, and A. Weil. The membership has varied over the years, sometimes involving as many as twenty mathematicians. The only rule of the group is to have no rules except compulsory retirement from membership at the age of fifty. The work of the group is based upon the unprovable metaphysical belief that for each mathematical question there is, among the many possible ways of dealing with it, a best, or optimal, way. In all likelihood the group adopted its pseudonym half in fun and half to avoid the tedium of listing a long bunch of authors on the title pages of their works.

251° *The origin of the name Bourbaki.* Though the founders of the Bourbaki group have purposely shrouded the origin of the name Nicolas Bourbaki in mystery, there are a couple of legends that endeavor to explain the choice.

There was a colorful officer, General Charles Denis Sauter Bourbaki, who achieved some fame in the Franco-Prussian War. In 1862, when he was forty-six, he was offered the Kingship of Greece, which he declined. In a disastrous military campaign in 1871 he was forced to retreat into Switzerland, where he was interned and tried to shoot himself. Apparently his attempt at suicide failed, for he lived to the good age of eighty-three. There is said to be a statue of him in Nancy, France, and this might be the connection between him and the later group of mathematicians, for several of the group were at one time or another associated with the University of Nancy. This explanation still leaves the "Nicolas" part of the name unresolved.

Another legend concerning the origin of the name Bourbaki is based upon a story that, about forty years ago, entering students at the École Normale Supérieure, where so many French mathematicians received their training, were exposed to a lecture by a distinguished visitor named Nicolas Bourbaki, who in reality was merely a disguised amateur actor or perhaps an upperclassman, skilled in straight-faced and seemingly plausible mathematical double talk.

252° *Bourbaki's application for membership.* Nicolas Bourbaki once applied for membership to the American Mathematical Society. The officials of the Society, regarding the application as a sophomoric joke, were not amused. The application was rejected and the cold suggestion was made that Bourbaki might apply for an institutional membership. But the dues for institutional membership are considerably higher than those for individual membership, and, moreover, Bourbaki did not wish to concede his nonexistence. Accordingly, the Society heard nothing more on the matter.

253° *Charge and countercharge.* In the *1947 Year Book* of the *Encyclopaedia Britannica* appeared a brief note about the Bourbaki group. The author of the note was Ralph P. Boas, then executive editor of *Mathematical Reviews*. Not long after the note's appearance, the editors of the *Britannica* received a hurt letter signed by N. Bourbaki protesting Boas's asservation that N. Bourbaki did not exist. The confusion of the editors and the discomfiture of Boas were finally cleared up by a letter from the American Mathematical Society, signed by the same secretary who had rejected Bourbaki's application for membership in the Society. As revenge, Bourbaki caused a rumor to be circulated that Boas did not exist. "Boas," said Bourbaki, "is the collective pseudonym of a group of young American mathematicians who act jointly as the editor of *Mathematical Reviews*."

254° *Correcting an error.* One of the prime movers of the Bourbaki group has been Jean Dieudonné, who has also served as Bourbaki's chief scribe. Since Dieudonné writes a considerable amount of mathematics under his own name, he has sometimes experienced difficulty in separating his private work from his work for Bourbaki. In one instance he managed to keep the record straight in a personally satisfying way. It seems that he once published a note, under Bourbaki's name, that was later found to contain an oversight. The oversight was corrected in a subsequent paper entitled "On an error of M. Bourbaki" and signed Jean Dieudonné.

255° *The model for Bourbaki.* It has been acknowledged by Jean Dieudonné that Van der Waerden's famous treatise on algebra served as the original model for the Bourbaki work. Indeed, the Van der Waerden book can be considered as Bourbachique, not only because of its style and spirit, but also since Van der Waerden in his preface says that the treatise really had several authors, including Emmy Noether and Emil Artin.

256° *A ball of tangled yarn.* The Bourbaki conception of present-day mathematics, or at least Jean Dieudonné's conception, is that mathematics today is like a ball of many tangled strands of yarn (see Figure 27), where those strands in the center of the ball react tightly upon one another in a nigh unpredictable manner. In this tangle of yarn there are strands, and ends of strands, that issue outward in

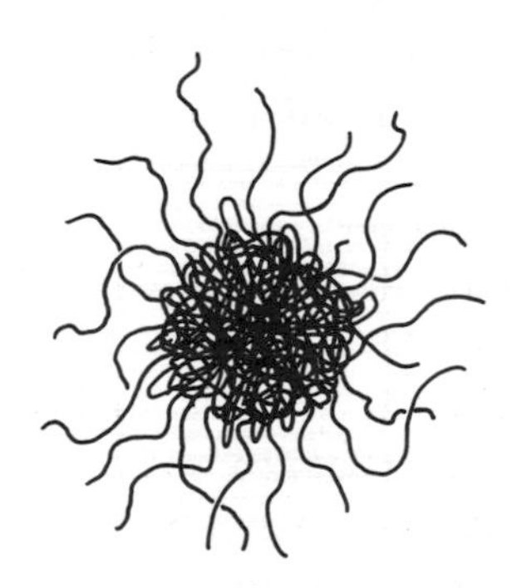

FIGURE 27

various directions and which have no intimate connection with anything within. The Bourbaki method is to snip off these free strands and to concentrate only on the tight core of the ball from which all the rest unravels. This tight core contains the basic structures and the fundamental processes or tools of mathematics—those parts of mathematics that have graduated from stratagems to methods, and have attained a goodly degree of fixedness. It is only this part of mathematics that Bourbaki attempts to arrange logically and to shape into a coherent and easily applied theory. It follows that much of mathematics is purposely left outside the province of the Bourbaki group.

257° *How Bourbaki works.* The Bourbaki method of work is long and painful. The group holds meetings two or three times a year, where the necessity of doing certain books or chapters is agreed upon. The job of forming a first draft is put in the hands of a willing member, and this person sets about the construction of his version based upon the rather vague plan of the group. After one or two years, when his work is complete, he submits it to the Bourbaki Congress, where each page and line is read aloud and mercilessly criticized by those present. The argument becomes vociferate, animated, and virulent, and the meeting seems to turn into a gathering of madmen. Once the first draft has been pitilessly shredded to pieces, a second willing collaborator starts the task over again. In turn, his version is torn apart, and a third man will start, and so on until for purely human reasons the process comes to an end. So, after everyone has seen the same book or chapter from six to a dozen times, there finally comes a unanimous vote to send it to the press. Even this edition is often improved upon in some subsequent edition of the work. It takes an average of eight to twelve years from the time the group first sets to work on a book or chapter to the time the work appears in the bookshops.

258° *Oil for the lamp.* Of considerable surprise to the Bourbaki group has been the commercial success, particularly in America, of their growing treatise. This success has proved to be very fortunate, for the resulting royalties have provided ample money for travel and for the fine foods and beverages that "lubricate" the periodic proceedings of the corporation.

259° *Worthy of imitation.* An interesting symbol, that deserves employment by other authors, has been introduced into the writings of the Bourbaki group. Whenever the mathematics becomes especially tricky or slippery, the reader is warned in the margin by a conspicuous Z-shaped curve (see Figure 28), indicating a "dangerous turn" in the argument.

FIGURE 28

The Bourbaki group has also favored simple terminology and has accordingly abolished much mathematical terminology used elsewhere. Thus *hypersphere* and *parallelotope* are replaced by *boule* (ball) and *pave* (paving-stone or brick).

260° *The University of Nancago.* There is a new series of advanced mathematics books being published under the impressive and confusing heading: *Publications de l'Institut Mathématique de l'Université de Nancago.* One cannot find a University of Nancago anywhere in the world, but the name takes on meaning when one realizes that the books of the series involve an author from the University of Nancy and an author from the University of Chicago.

ARCHIMEDES TO SIDNEY CABIN

261° *Archimedes' tomb found.* Apropos Item 79° of *In Mathematical Circles*, is the following article released by the New York Times News Service in 1965:

> ROME—The tomb of Archimedes in which his ashes are interred has been found after 2,200 years, during excavation for the foundations of a hotel in Syracuse, Southwest Sicily, it was

reported yesterday by the Italian news agency ANSA.

For 150 years it was believed that the tomb of the ancient Greek mathematician was the one found near the Greek theater in Syracuse.

ANSA said the ashes of Archimedes were found in a lead case about six feet long studded with gems and gold seals. The case was under two burial stones held together by lead-coated bars in the center of a stone platform that covers an area of more than 80 square yards.

According to Prof. Salvatore Ciancio, honorary inspector of antiquities, Syracuse, the tomb corresponds to a description given by Cicero, ANSA said.

HUBERT C. KENNEDY.

262° *A sizeable breakfast.* One of the most amusing anecdotes involving trouble with a foreign language is that told of Charles Babbage and John Herschel when they were on a visit to France. At breakfast one morning, Babbage ordered two eggs for each of them by telling the waiter, "pour chacun deux." Thereupon the waiter called out to the kitchen, "Il faut faire bouiller cinquante deux oeufs pour Messieurs les Anglais." The alarmed breakfasters fortunately succeeded in stopping the cook in time.

263° *War and peace.* There is so much to admire in the life of Benjamin Banneker (1731–1806). He was the first American Negro mathematician; he published a very meritorious almanac from 1792 to 1806, making his own astronomical calculations; using a borrowed watch as a model, he constructed entirely from hard wood a clock that served as a reliable timepiece for over twenty years; he won the enthusiastic praise of Thomas Jefferson, who was then the Secretary of State; he served as a surveyor on the Commission appointed to determine the boundaries of the District of Columbia; he was known far and wide for his ability in solving difficult arithmetical problems and mathematical puzzles quickly and accurately. These achievements are all the more remarkable in that he had almost no formal schooling and was therefore largely self-taught, studying his mathematics and astronomy from borrowed books while he worked for a living as a farmer.

But laudable as all the accomplishments of Benjamin Banneker mentioned above are, there is a further item that perhaps draws stronger applause. In his almanac of 1793, he included a proposal for the establishment of the office of Secretary of Peace in the President's Cabinet, and laid out an idealistic pacifist plan to insure national peace. Every country in the world has the equivalent of a Secretary of War. Had Benjamin Banneker's proposal been sufficiently heeded, the United States of America might have been the first country to have a Secretary of Peace! The possibility of realizing this honor still exists—and the time for it is overripe.

264° *A Benjamin Banneker problem.* Benjamin Banneker received challenging mathematical problems from scholars in various parts of the country who seemed interested in testing his ability. It is said that he never failed to return a solution. He propounded questions of his own for others to solve, frequently presenting them in rhymes of his own composition. Following is one of Benjamin Banneker's rhyming problems: perhaps an interested reader may care to wrestle with it.

A cooper and vintner sat down for a talk,
Both being so groggy that neither could walk;
Says cooper to vintner, "I'm the first of my trade,
There's no kind of vessel but what I have made,
And of any shape, sir, just what you will,
And of any size, sir, from a tun to a gill."
"Then," says the vintner, "you're the man for me.
Make me a vessel, if we can agree.
The top and bottom diameters define,
To bear that proportion of fifteen to nine,
Thirty-five inches are just what I crave,
No more and no less in the depth will I have;
Just thirty-nine gallons this vessel must hold,
Then I will reward you with silver or gold,—
Give me your promise, my honest old friend."

"I'll make it tomorrow, that you may depend!"
So, the next day, the cooper, his work to discharge,
Soon made the new vessel, but made it too large;
He took out some staves, which made it too small,
And then cursed the vessel, the vintner, and all.
He beat on his breast, "By the powers," he swore
He would never work at his trade any more.
Now, my worthy friend, find out if you can,
The vessel's dimensions, and comfort the man!

BENJAMIN BANNEKER

265° *Great grandson of an African king.* Benjamin Banneker was born on November 9, 1731, on a farm close to Baltimore, Maryland. His mother was a free Negro and his father was a slave. His maternal grandmother was an English white woman legally married to a native African who was the son of an African king.

In August of 1791, Banneker wrote to Thomas Jefferson his celebrated letter pleading for the rights of the thousands of slaves held in the Colonies. Along with his letter he sent a copy of his forthcoming 1792 almanac. Jefferson, who was then Secretary of State, replied to Banneker in part as follows:

> Nobody wishes more than I do to see such proofs as you exhibit that Nature has given to our black brethren talents equal to those of the other colors of men, and that the appearance of a want of them is owing only to the degraded condition of their existence both in Africa and America. . . . I have taken the liberty of sending your almanac to M. de Condorcet, secretary of the Academy of Sciences at Paris, and member of the Philanthropic Society, because I consider it a document to which your whole color has a right for their justification against the doubts which have been entertained of them.

266° *A good reason.* I once attended a series of lectures on the theory of approximation, given by Besicovitch. In the first lecture Besicovitch pointed out with great care that "there is no T in the name Chebyshoff." In the second lecture he introduced T-polynomials. "We call them T-polynomials," he explained, "because T is the first letter of the name Chebyshoff."—RALPH P. BOAS

267° *Another good reason.* When lecturing before an analytic geometry class during the early part of the course, one may say: "We designate the slope of a line by *m*, because the word *slope* starts with the letter *m*; I know no better reason."

268° *A paper in "The Physical Review."* Hans Bethe and George Gamow some years ago became acquainted with an upcoming bright young physicist possessing an unusual name. The result was the publication by the three men of a paper that would not have been conceived had their names been different. On April 1 (note the date), 1948, there appeared, in *The Physical Review*, a perfectly straight-faced paper on the origin of chemical elements. The only unusual feature of the paper was the authorship by-line, which read: "by Alpher, Bethe, and Gamow."

269° *A commemoration.* Mathematical research in Hungary began with Farkas (Wolfgang) Bolyai and his son János (Johann) Bolyai. The Bolyai family is memorialized in Budapest today in the names of three of the city's streets.

270° *Well integrated person.* When Sidney Cabin taught mathematics at The Cooper Union School of Engineering, he used to ask his students to integrate d(Cabin)/Cabin. (Neglect the constant of integration.)—ALAN WAYNE

QUADRANT FOUR

From an early recognition
to Wiener's famous letter

CAUCHY TO COOLIDGE

271° *Cauchy's early recognition.* On the first of January, 1800, Cauchy's father was elected Secretary of the Senate and was furnished with an office in the Luxembourg Palace. Cauchy, who at the time was only eleven, shared his father's office with a small desk in a corner. Lagrange, then a professor at the École Polytechnique, frequently dropped in to discuss business with Secretary Cauchy, and in this way became acquainted with the boy's mathematical talent. One day, in the presence of a number of notables, pointing to the boy, Lagrange said, "Some day that little man will supersede all of us, insofar as we are mathematicians."

272° *Cauchy's productivity.* Cauchy produced mathematics at so great a rate that he was forced to found a sort of journal of his own, the *Exercises de Mathématiques* (1826–1830), which was followed by a second series called *Exercises d'Analyse Mathématique et de Physique*, into which he poured both his expository and original works in mathematics. Mathematicians eagerly bought and studied these works, and they became quite influential.

An amusing story is told in connection with Cauchy's prodigious productivity. In 1835 the Academy of Sciences began publishing its *Comptes Rendus*. So rapidly did Cauchy supply this journal with articles that the Academy became alarmed over the mounting printing bill, and accordingly passed a rule, still in force today, limiting all published papers to a maximum length of four pages. Cauchy had to seek other outlets for his longer papers, some of which exceeded one hundred pages.

Cauchy's total output contains 789 papers, some of which are very extensive works, and they fill twenty-four large quarto volumes! He has been criticized for overproduction and overhasty composition.

273° *Cauchy's bigotry.* In religion Cauchy was bigoted and quite intolerant of another's creed. He spent much of his time trying to convert others to his particular belief. When William Thomson

(Lord Kelvin), then only twenty-one years old, visited Cauchy to discuss mathematics, Cauchy spent the entire time trying to convert the young man—who was a staunch adherent of the Scottish Free Church—to Catholicism.

Cauchy was frequently accused of voting new members into scientific societies in accordance with his personal religious and political views rather than on the candidates' scientific merits. This led Cauchy to become unpopular with many of his colleagues.

274° *Cauchy and the oath of allegiance.* The case of Cauchy and the oath of allegiance is apropos in these days when scholars are frequently harassed with loyalty oaths. Cauchy's trouble started in 1830, when a revolution unseated King Charles X in favor of Louis Philippe. Cauchy, who refused to swear an oath of allegiance to the new government, had to leave his chair in the French Academy and go into exile, first in Switzerland, then Turin, and finally Prague.

In 1838, when he was about fifty, Cauchy returned to France and again assumed his seat in the French Academy, for a special dispensation had been passed exempting members from taking the oath of allegiance to the government. At this time Cauchy was unanimously elected to fill a vacancy at the Collège de France, but this he had to forego, for again he ran into the oath of allegiance, which he refused to take. He was then unanimously elected to the Bureau des Longitudes, and his presence was so sought there that the government finally gave in to the Bureau and relinquished the oath of allegiance insofar as Cauchy was concerned. But the government tried to get the Bureau to elect someonc else to fill Cauchy's position, which he refused to vacate. Finally, in 1843, Cauchy publicized his case in the form of an open letter to the people. His letter was a magnificent defense of freedom of conscience and thought. At length—through riots, strikes, and civil war—the stupidity of repression was brought home to the government. In 1848 Louis Philippe was ousted, and one of the first acts of the succeeding Provisional Government was to abolish the oath of allegiance.

In 1852, when Napoleon III founded the Second Empire, the oath was restored, but it was quietly acknowledged that Cauchy could continue his lectures without taking it.

275° *Cauchy's last words.* Cauchy died suddenly on May 23, 1857, when he was sixty-eight years old. He had gone to the country to rest and to cure a bronchial trouble, only to be smitten by a fatal fever. Just before his death he was talking with the Archbishop of Paris. His last words, addressed to the Archbishop, were: "Men pass away, but their deeds abide."

276° *Cayley and Sylvester on Euclid.* J. J. Sylvester, who ardently supported reform in the teaching of geometry in England, once said that he wished to bury Euclid "deeper than e'er plummet sounded" out of the reach of the schoolboy. On the other hand, Arthur Cayley, who was a fervent admirer of Euclid, said that he desired the retention in the English schools of Simson's *Euclid.* When he was reminded that this treatise is not pure Euclid, but is a mixture of Euclid and Simson, he suggested striking out the Simson contributions and sticking strictly to the original Euclid.

277° *Cayley and Tait on quaternions.* S. P. Thomson, in his *Life of Lord Kelvin* (1910), tells an amusing anecdote concerning Peter Guthrie Tait and Arthur Cayley. Tait was an avid champion of Hamilton's quaternions; Cayley, on the other hand, found no use for them. In urging quaternions on Cayley, Tait said, "You know quaternions are just like a pocket-map." "That may be," replied Cayley, "but you've got to take it our of your pocket, and unfold it [into Cartesian coordinates] before it's of any use." And he smilingly dismissed the matter.

278° *Some of William Kingdon Clifford's personal characteristics.* Clifford's code of ethics involved freedom and independence: "There is one thing in the world more wicked than the desire to command, and that is the will to obey." He was free of hypocrisy; a letter he wrote to Lady Pollock on his theory of ideal behavior concluded with: "All this, by the way, is only theory; my practice is just like other people's." He disliked pretentiousness; of an acquaintance contemplating writing a work on philosophy he humorously observed: "He is writing a book on metaphysics, and is really cut out for it; the clearness with which he thinks he understands things and his total inability to express what little he knows will make his fortune as a philosopher." He was incapable of personal enmity and once said: "I believe if all the murderers and all the priests and all the liars in the world were united into one man, and he came suddenly upon me round a corner and said, 'How do you do?' in a smiling way, I could not be rude to him upon the instant."

279° *Professor Coolidge and examinations.* Professor Julian Lowell Coolidge, the great geometer of Harvard University, used to pace around the room while one of his classes was taking an examination, and glance at the various students' efforts as he walked. If his eye fell on something which displeased him he would point it out to the student and advise him to work the problem over again.—JOHN K. MOULTON

280° *Professor Coolidge's test.* At a large mathematical gathering, Professor Coolidge rose, advanced to the front of the room, and proceeded to frighten the group by announcing that he was going to give the group a little mathematics test. Now professors of mathematics may like to give tests, but taking one is another matter. To calm the audience, he said he merely wanted to verify that most mathematicians know very little elementary solid geometry.

Professor Coolidge first reviewed a few definitions, such as that of *median* and of *altitude* applied to triangles and tetrahedra. "Now," he asked, "though, as any high school student of geometry knows, the medians of a triangle are concurrent, can the same be said of the medians of a tetrahedron?" After some hesitation, almost everyone present said that surely they must be. Professor Coolidge assured them that this is indeed the case. He next similarly asked, "Though, as any high school student of geometry knows, the altitudes of a triangle are concurrent, can the same be said of the altitudes of a tetrahedron?" Many present immediately said that of course they are concurrent, most of the others said they blame well ought to be, and the few remaining ones, fearing some sort of a trap, were noncommittal. Professor Coolidge then explained that the altitudes of a tetrahedron usually are not concurrent, and that concurrency occurs only in the so-called *orthocentric tetrahedra*, in which each edge of the tetrahedron is perpendicular in space to the opposite edge of the tetrahedron.

Professor Coolidge went on to other simple questions and well verified his thesis that most mathematicians know very little elementary solid geometry.

DEDEKIND TO GERBERT

281° *An incorrect obituary.* Heinrich Tietze has told an amusing story about Richard Dedekind. It seems that in 1904 an academic calendar appeared, with the motto *nulle dies nisi festiva,* which recorded a birth date or a death date of some mathematician for each day of the year. On it, September 4, 1899, was marked as the day of Dedekind's death. Dedekind wrote to the publisher of the calendar, saying that though September 4 might be correct, 1899 certainly was not, for on that day he was in the best of health and had enjoyed a stimulating mathematical discussion with his dinner guest and honored friend, Georg Cantor of Halle, who, Dedekind added, used the opportunity to deal a death blow not to his friend but to a mathematical error he had made. The calendar turned out to be completely wrong, for Dedekind died on February 12, 1916.

282° *Competition for Professor Lee Swinford.* At a departmental picnic in Jackson Park, Chicago, Professor Leonard E. Dickson entertained the group with some of his corny puns.

What geometric construction reminds you of a dog in the refrigerator?" Answer: "A purp in the cooler" (a perpendicular).

"How does a piece of writing paper remind you of a lazy dog?" Answer: "A piece of writing paper is an ink-lined plane, and an inclined plane is a slope up, and a slow pup is a lazy dog."

I. A. BARNETT

283 *Dido's problem.* After the murder of her husband by her brother Pygmalion, Dido set sail with all her wealth and some faithful friends, and landed, or was shipwrecked, on the coast of North Africa. She asked the ruler of the area for as much territory as could be encompassed by the hide of a bull, as asylum for herself and her friends. When the seemingly modest request was granted, she cut a bull's hide into the thinnest possible strips, fastened these together into a long belt, and ran the belt from one point on the coast to another, thus gaining the coastal area bounded by the belt. This is the legend that tells how Carthage was founded.

The above legend gave rise to the mathematical problem, known as Dido's problem, of finding the maximum area that can be encompassed by a curve of given length. This, in turn, gave birth to the field of mathematics now known as the calculus of variations.

284° *Related to Dido's problem.* There is a cunning little trick wherein a person takes a piece of paper, say a piece of typing paper, and boasts that he can cut a clever hole in it through which he can push a full-sized chair. The boast, of course, elicits exclamations of skepticism from the audience, and so, before his incredulous viewers, the trickster cuts a hole in the paper through which he can stick his arm, and then with this arm he pushes a full-sized chair across the room.

There actually is, however, a suitable hole that can be cut in the piece of paper and through which (theoretically, at any rate) a full-sized chair can be pushed in the fashion originally imagined by the trickster's audience. (See, for example, Figure 29.)

285° *The sad case of Maurice de Duffahel.* Maurice de Duffahel achieved a temporary fame by simply republishing, under his own name, a number of earlier classical papers by some of the great mathematical masters. He made practically no attempt to disguise these papers. In 1936 he fraudulently published in this way a paper that twenty-four years earlier had been published by Charles Émile Picard. Duffahel's paper was absolutely identical in words and symbols with the Picard paper, except that Duffahel found it necessary to omit one footnote in whch the original author had referred to one of his previous papers. A reviewer of the Picard-Duffahel paper was well enough acquainted with the works of Picard to recognize the duplicity, and Duffahel's publishing career came to a sudden end. As Paul R. Halmos remarked, the moral appears to be: "You can fool some editors some of the time, but you can't fool all reviewers all of the time."

286° *Expert extraordinary.* Samuel Eilenberg, originally from Warsaw, Poland, and now a Professor of Mathematics at Columbia

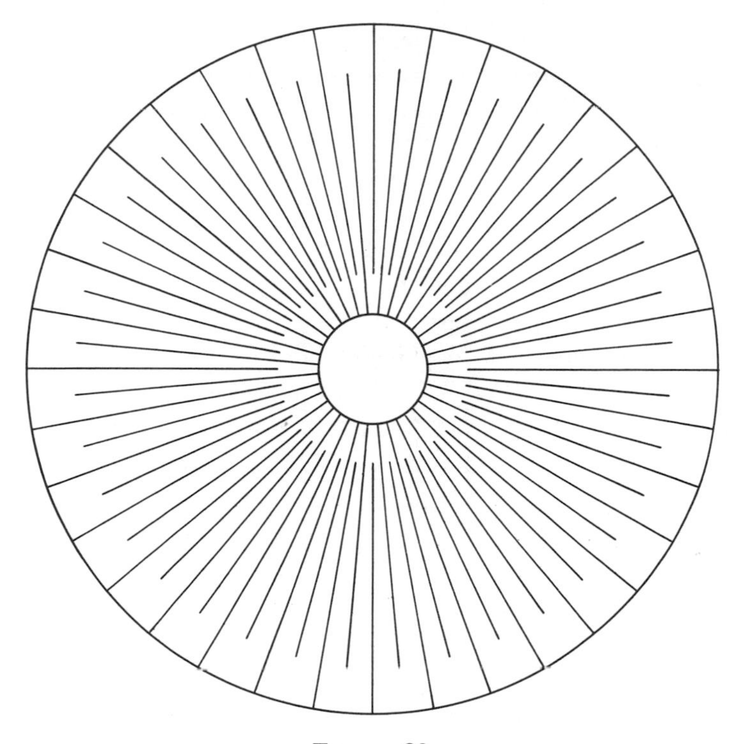

FIGURE 29

University, is an expert extraordinary in algebraic topology. His early mathematical brilliance was so great that friends of his youth affectionately called him S^2P^2—for Smart Sammy the Polish Prodigy.

287° *Dire extremity.* In an address at Syracuse, May 11, 1957, Professor Howard F. Fehr of Columbia University Teachers College remarked, "A mathematics professor who talks at length affects both ends of the listener—he makes one end feel numb and the other feel dumb."—ALAN WAYNE

288° *Fejér's wit.* [This and the next item are adapted, with permission, from the entertaining article "Some mathematicians I have known," by George Pólya, which appeared in *The American Mathematical Monthly*, August-September 1969, pp. 746–753.]

It happened at a meeting in Germany. At that time I was a

"Privatdozent." I cannot completely explain what that is: A financially shaky position, somewhat similar to, but not quite, an Assistant Professor—thank goodness, this institution of Privatdozents has started to disappear nowadays. I was married and my wife took photographs of the mathematicians. She stopped Fejér [Lipot Fejér (1880–1959)] in the company of three or four others, in front of the university on the street car tracks, took a picture and was about to take a second one as Fejér spoke up. "What a good wife! She puts all these full professors on the tracks of the street car so that they may be run over and then her husband will get a job!"—GEORGE PÓLYA

289° *More of Fejér's wit.* At another meeting, several years later, Fejér was very angry (and with some good reason) at a Hungarian mathematician, a topologist whose name I shall not tell you. I walked up and down a long time with Fejér, who could not stop talking about the target of his anger and wound up by saying: "And what he says is a topological map of the truth." You must realize how distorted a topological map may be.—GEORGE PÓLYA

290° *The Gerbert riddle.* There is a riddle concerning the famous French scholar, mathematician, and churchman Gerbert (ca. 950–1003): *Scandit ab R Gerbertius in R, post Papa viget R* (From R Gerbert ascended to R, and then reached the summit as Pope at R).

[Gerbert was appointed Archbishop of *Rheims* in 991; he was elevated to Archbishop of *Ravenna* in 998; he became Pope Sylvester II at *Rome* in 999.]

HAMILTON AND HARDY

HAMILTON and Hardy were alphabetically linked together in a section of *In Mathematical Circles*, and here they are similarly linked together again. There are many anecdotes about these two mathematicians. The five stories below on Hardy by George Pólya are adapted, with permission, from Pólya's article entitled "Some mathematicians I have known," which appeared in the August-September 1969 issue of *The American Mathematical Monthly*, pp. 746–753. George Pólya, in addition to being perhaps the foremost combination in the United States of outstanding researcher and inspirational teacher of mathematics, is a gifted raconteur of mathematical stories and anecdotes.

291° *How Hamilton was taken.* Sir Edmund Whittaker, who in 1906 was appointed to the chair once held by Sir William Rowan Hamilton, has told an amusing anecdote about Hamilton that was reported by people of the countryside. It seems that one of Hamilton's duties as Royal Astronomer was to administer the 17 acres of farmland around Dunsink Observatory. Now Hamilton, who was town-bred, knew little of farming, but to supply his household with milk he purchased a cow. When, in the natural course of events, the milk yield fell off, Hamilton consulted a neighboring farmer. The farmer, well aware of whom he was dealing, said that the cow, being the only occupant of the 17 acres, was suffering from loneliness. Hamilton inquired if it would be possible to provide the cow with companions, and the farmer, for a recompense, graciously agreed to allow his own cattle to graze on the rich pastureland of Dunsink.

292° *How Hamilton became interested in mathematics.* William Rowan Hamilton was not brought up by his parents. When the youngster was about a year old it was decided to entrust his education to his father's brother James, a clergyman who had settled in a small village about 30 miles north of Dublin. The uncle gave the boy a strenuous and lopsided education.

William proved to be a prodigy. He could read English at the age of three; at four he was able to translate Latin, Greek, and Hebrew; at eight he added Italian and French. Before he was ten he was a student of Arabic and Sanskrit, and at fourteen he composed a poem in Persian addressed to the Persian Ambassador when that official visited Dublin. The boy was fond of the Classics and of poetry.

It was not until he was fifteen that William's interests changed and he became excited about mathematics. This change was brought about by his meeting Zerah Colburn, the American lightning calculator (see Item 300° of *In Mathematical Circles*), who, then only a youngster himself, gave a demonstration of his powers at an exhibition in Dublin.

For a long time after that, William practiced mental calculation, and he resolved to make mathematics his prime concern.

293° *Hamilton's alarm clock.* When Hamilton was about sixteen, his uncle would awake him at five in the morning by pulling a string fastened to the boy's nightshirt and passing through the wall between their bedrooms.

294° *A Hamilton quote.* Edgar Allen Poe spoke of "The glory that was Greece and the grandeur that was Rome." In the same vein Hamilton asked: "Who would not rather have the fame of Archimedes than that of his conqueror Marcellus?"

295° *Hardy and the Riemann hypothesis.* Hardy wrote very well and with great facility, but his papers, especially some of his joint papers with Littlewood, make no easy reading. The problems are very hard and the methods unavoidably very complex. He valued clarity, yet what he valued most in mathematics was not clarity but power, surmounting great obstacles that others abandoned in despair. He himself had very great power, and he was fascinated by the Riemann hypothesis.*

Hardy loved sunshine, but there is not much sunshine in England. Therefore, in the summer vacation he went regularly to the continent as soon as the cricket season was over, and visited with friends. His principal friend was Harald Bohr. They had set a routine. First they

---

* The Riemann hypothesis is a celebrated unproved conjecture which is to classical analysis what Fermat's "last theorem" is to number theory. Euler had pointed out connections between the theory of prime numbers and the series

$$1/1^s + 1/2^s + 1/3^s + \cdots + 1/n^s + \cdots,$$

where s is an integer. Riemann studied the same series for s a complex number $\sigma + i\tau$. The sum of the series defines a function $\zeta(s)$ which has come to be known as *Riemann's zeta function.* Riemann, around 1859, conjectured that all the imaginary zeros of the zeta function have their real part $\sigma = 1/2$. In 1914 Hardy succeeded in showing that $\zeta(s)$ has an infinity of zeros with $\sigma = 1/2$, but, though now over a century old, the original Riemann conjecture is still unresolved.

sat down and talked, and then they went for a walk. As they sat down, they made up and wrote down an agenda. The first point of the agenda was always the same: "Prove the Riemann hypothesis." Of course, this point was never carried out. Still, Hardy insisted that it should be written down each time.—GEORGE PÓLYA

296° *Hardy criticizes Pólya.* In working with Hardy, I once had an idea of which he approved. But afterwards I did not work sufficiently hard to carry out the idea, and Hardy disapproved. He did not tell me so, of course, yet it came out when he visited a zoological garden in Sweden with Marcel Riesz. In a cage there was a bear. The cage had a gate, and on the gate there was a lock. The bear sniffed at the lock, hit it with his paw, then he growled a little, turned around and walked away. "He is like Pólya," said Hardy. "He has excellent ideas, but does not carry them out."—GEORGE PÓLYA

297° *Hardy and sports.* Hardy had very personal and very definite views about all sorts of things. He liked cats, but could not stand dogs. He loved cricket, but despised rowing.

Now Hardy was a Cambridge man, but for some time he was professor at Oxford. It happened that at Oxford somebody who was unaware of Hardy's idiosyncrasies asked him: "For which university are you in sports?" "It depends," said Hardy. "In cricket I am for Cambridge, in rowing I am for Oxford."—GEORGE PÓLYA

298° *Hardy tries to outsmart God.* Hardy believed God was his personal enemy. You understand: God has nothing more urgent to do than to annoy Hardy. There is a multitude of stories concerning the contest of wits between God and Hardy. Here is the best known.

Hardy stayed in Denmark with Bohr until the very end of the summer vacation, and when he was obliged to return to England to start his lectures there was only a very small boat available (there was no airplane traffic at that time). The North Sea can be pretty rough and the probability that such a small boat would sink was not exactly zero. Still, Hardy took the boat, but sent a postcard to Bohr: "I proved the Riemann hypothesis. G. H. Hardy."

You must understand the underlying theory: If the boat sinks and Hardy drowns, everybody must believe that he proved the Riemann hypothesis; but God would not let Hardy have such a great honor and so He will not let the boat sink.—GEORGE PÓLYA

299° *Another attempt by Hardy to outsmart God.* Another summer Hardy stayed in Engelberg, an Alpine valley in Switzerland where we had a chalet. He liked the sunshine but it rained all the time, and as there was nothing else to do, we played bridge: Hardy, who was quite a good bridge player, my wife, myself, and a friend of mine, F. Gonseth, mathematician and philosopher. After a while Gonseth had to leave; he had to catch a train. I was present as Hardy said to Gonseth: "Please, when the train starts, open the window, stick your head through the window, look up to the sky, and say in a loud voice, 'I am Hardy.'"

Again, you must understand the underlying theory: When God thinks that Hardy has left, He will make good weather just to annoy Hardy.—GEORGE PÓLYA

300° *Hardy and mirrors.* One of Hardy's idiosyncrasies was an intense dislike of mirrors. All mirrors were barred from his living quarters, and the first thing he did upon checking into a hotel suite was to reverse or cover all mirrors in the rooms.

301° *International understanding.* At a meeting of mathematicians in France, the English mathematician J. E. Littlewood was greeted by a French mathematician: "So there really is a Littlewood, and it is not just a pseudonym which G. H. Hardy uses to publish his poorer papers!"—ALAN WAYNE

HEILBRONN TO HURWITZ

302° *On age.* I was having dinner with Heilbronn at Trinity, and the conversation turned to wines. I remarked that I took some pleasure in having been born in a very good year for port. Heilbronn thought for a barely perceptible instant, and then said, "Why, I thought you were older than that."—RALPH P. BOAS

303° *An unlikely thought.* Professor David Hilbert came limping through the snow on one snowshoe, holding a broken one in his arms. When asked why he had not removed the other one, he said that that had not occurred to him.—I. A. BARNETT

304° *Hilbert and the Riemann hypothesis.* [This and the next two items are adapted, with permission, from George Pólya's article "Some mathematicians I have known," which appeared in the August-September 1969 issue of *The American Mathematical Monthly*, pp. 746–753.]

There is a German legend about Barbarossa, the emperor Frederick I. The common people of Germany liked him and as he died in a crusade and was buried in a faraway grave, the legend sprang up that he was still alive, asleep in a cavern of the Kyffäuser mountain, but would awake and come out, even after hundreds of years, when Germany needed him.

Somebody allegedly asked Hilbert, "If you could revive, like Barbarossa, after five hundred years, what would you do?" "I would ask," said Hilbert, "Has somebody proved the Riemann hypothesis?"

GEORGE PÓLYA

305° *Hilbert's absent-mindedness.* There was a party at Hilbert's house and Frau Hilbert suddenly noticed that her husband forgot to put on a fresh shirt. "David," she said sternly, "go upstairs and put on another shirt." David, as it befits a long married man, meekly obeyed and went upstairs. Yet he did not come back. Five minutes passed, ten minutes passed, yet David failed to appear and so Frau Hilbert went up to the bedroom and there found Hilbert asleep in bed. You see, it was the natural sequence of things: He took off his coat, then his tie, then his shirt, and so on, and went to sleep. [For a parallel Newton story, see Item 194° of *In Mathematical Circles*.]

GEORGE PÓLYA

306° *Hilbert receives a new professor.* A new member of the Göttingen faculty was supposed to introduce himself formally to his colleagues. He put on a black coat and a top hat, took a taxi, and made the round of the faculty houses. The taxi stopped in front of each, and the new colleague presented his visiting card at the door. Sometimes he got the answer that the Herr Professor is not at home, but when the Herr Professor was at home the new colleague was supposed to go in and chat for a few minutes. Once such a new colleague came to Hilbert's house and Hilbert decided (or Frau Hilbert decided for him) that he was at home. So the new colleague came in, sat down, put his top hat on the floor, and started talking. This was the proper thing to do, but he did not stop talking. And Hilbert—the visit probably interrupted some mathematical meditation—became more and more impatient. And what did he finally do? He stood up, took the top hat from the floor, put it on his head, touched the arm of his wife, and said: "I think, my dear, we have delayed Herr Kollege long enough."—and walked out of his own house.—GEORGE PÓLYA

307° *Euclid's "Elements" and Hilbert's dictum.* The origin of early Greek mathematics is clouded by the greatness of Euclid's *Elements*, written about 300 B.C., because this work so clearly excelled so many preceding Greek writings on mathematics that the earlier works were thenceforth discarded and have become lost to us. As David Hilbert (1862–1943) once asserted, "One can measure the importance of a scientific work by the number of earlier publications rendered superfluous by it."

308° *Got his number.* At a Joint Luncheon and Panels of the Association of Teachers of Mathematics and the Mathematics Chairmen's Association (both of New York City), held at Columbia University, Julius Hlavaty told the following story to a group of educators among whom was John R. Clark of Teachers College, Columbia University:

"My four-year-old son proudly brought me a copy of Professor Clark's elementary arithmetic textbook. The boy had completed an exercise in addition involving the sum of two integers, and to my amazement, more than half of the answers were correct. My son (a good reader, but innocent of numbers, except for counting) said the exercise was easy. The exercise was entitled, 'Additions Involving 4.' Said he, 'Wherever I saw there was no 4 in the problem, I put 4 in the answer!'"—ALAN WAYNE

309° *The aphoristician.* [This and the following item are adapted, with permission, from George Pólya's article "Some mathematicians I have known," which appeared in the August-September 1969 issue of *The American Mathematical Monthly*, pp. 746–753.]

Adolph Hurwitz (1859–1919) was very much like Fejér in the style of his work. Felix Klein, in his *History of Mathematics in the Nineteenth Century*, calls Hurwitz an "aphoristician." An aphorism is a concise weighty saying. The aphorism is short, but its author may have worked a long time to make it so short. Hurwitz's papers are like aphorisms. In the wide range of his mathematical knowledge he spotted well circumscribed weighty problems capable of a surprisingly simple solution and presented the solution in perfect form. If you wish to see an easily accessible sample, read two pages of his collected works—the proof for the transcendence of the number *e*.—GEORGE PÓLYA

310° *Pushed too far.* I never heard Hurwitz utter a sharp sentence in public. Yet in the circle of his family or with good friends he could find a sharp or witty word. I must preface a little what I wish to quote. In discharging conscientiously his duties as a professor, he took care of many Ph.D. candidates, treating them with much consideration and patience. Among so many there were some who needed a lot of help, and even the patient Hurwitz was once led to say: "A Ph.D. dissertation is a paper of the professor written under aggravating circumstances."—GEORGE PÓLYA

KASNER TO LAWRENCE

311° *Student activity.* At a joint luncheon of the teachers and chairmen of mathematics of New York City, Edward Kasner, then Adrian Professor of Mathematics at Columbia University, related the following anecdote concerning Robert Adrian:

"When Adrian came to Columbia from Rutgers, although noted for his scholarship, he found the same discipline problems (including paper throwing) at Columbia as at Rutgers. When asked how the students at Columbia compared with those at Rutgers, Adrian remarked, drily, that the Columbia students appeared to be superior in aim."—ALAN WAYNE

312° *Proper subset.* Among the most attractive publishing offices in America are those of Simon & Schuster, at Rockefeller Center in New York City. When they first opened the offices, Professor Kasner, the mathematician, was taken on a tour of the premises. "This is the most modern office in the world," said Schuster. Kasner said he didn't believe it. "The most modern office in the whole civilized world," Schuster insisted. "Oh, in the *civilized* world," said Professor Kasner. "That narrows the field considerably."—ALAN WAYNE

313° *Hopeful attitude.* One day a student in Professor Edward Kasner's class was writing a proof on the blackboard and had to kneel to get the last lines written. "That's what I have always thought," remarked Professor Kasner. "There *is* an element of prayer in mathematics."—JEROME MANHEIM

314° *A self reference.* The story of Kerékjártó's book on topology with the picture of Bessel-Hagen is well known. It is less well known that (although the book contains many things originated by Kerékjártó himself) there is only one reference in the index to Kerékjártó. If you look it up, you find no mention of Kerékjártó, but there is a footnote: "Diesen falschen Satz habe ich bewiesen."—RALPH P. BOAS

315° *Transformation.* At Columbia University, Professor Ellis R. Kolchin was lecturing to a class and had begun a description of "partial order," "simple order," "complete order," and so on, for about ten kinds of "order." A student entered the room late and asked his neighbor as he sat down, "What's Kolchin doing?"

"He's making chaos out of order," was the reply.—ALAN WAYNE

316° *Lagrange's dictum.* Lagrange once remarked that a mathematician has not thoroughly understood a piece of his own work until he has so clarified it that he can effectively explain it to the first man he meets in the street. Though this ideal often appears impossible, time frequently renders it attainable. Newton's law of universal gravitation, which at first was incomprehensible to even highly educated persons, has today become commonplace. Einstein's relativistic theory of gravitation is now undergoing a similar transmutation.

317° *A celestial surprise.* Lagrange obtained some particular solutions of the three-body problem. One of these assumes that the three bodies start from the vertices of an equilateral triangle, in which case Lagrange showed that they then continue to move as though attached to the triangle, while the triangle rotates about the center of mass of the three bodies. It would seem that this particular case would have no physical reality, but in 1906 the case was found to apply to the sun, Jupiter, and an asteriod named Achilles.

318° *Consistent or inconsistent?* Although Laplace was universally admired for his mathematical and scientific genius, his political adaptability inspired widespread distrust. He was declared to be an opportunist and to be inconsistent in his loyalties. The stock appraisal became a comparison of him with the Vicar of Bray, for the Vicar was a supple man who was twice a Papist and twice a Protestant. But the Vicar defended himself from charges of accommodating himself to circumstances by replying: "Not so, for though I have changed my religions, I have kept true to my basic principle, which is to live and die as the Vicar of Bray."

319° *Laplace's last words.* Laplace died in 1827 and his last words have been reported to be: "What we know is very slight; what we don't know is immense." De Morgan, however, has reported Laplace's last words to be: "Man follows only phantoms."

320° *Mathematics as a science.* At Bates College, in Lewiston, Maine, during a meeting of the science and mathematics faculty, Dr. Walter Lawrence of the Chemistry Department said, "Now the sciences of physics, chemistry, and biology—and mathematics, if it can be considered a science...." At this point the Chairman of the Department of Mathematics interrupted and said, "Dr. Lawrence, just because mathematics doesn't stink and boil over is no reason not to consider mathematics as a science."

MILLER TO NEWTON

321° *Point well taken.* In the differential equations course at The Cooper Union, Professor Frederic H. Miller was discussing the deflection of steel beams according to loading. A student inquired, "If you have a problem in which the point falls outside the range of values, what happens?"

"It simply shows that you're off the beam," was the answer.

ALAN WAYNE

322° *A tour de force.* Purely as a challenging feat, Professor C. N. Mills, of Sioux Falls College, decided to calculate the radii of the eight circles that touch three given circles, using straightforward rectangular Cartesian coordinates. He finished the task in February, 1961. The complete solution, with all the work sheets, covered a strip of 18-inch wall paper 24 feet long!

One is reminded of a similar feat when, about 1894, a Professor Hermes of Lingen gave up ten years of his life to the problem of constructing a regular polygon of 65,537 sides (see Item 178° of *In Mathematical Circles*).

323° *The problem of neighboring domains.* The problem of neighboring domains came up in a discussion between August Ferdinand Möbius and his friend Adolph Weiske. It perhaps originated with Weiske, who, though not a mathematician by profession, had a keen interest in mathematical problems and puzzles. Möbius first presented the problem in a lecture in 1840; he offered it in the form of the following fairy tale.

Once upon a time in the Far East, there lived a King who had five sons, who were to inherit the kingdom after the King died. In his will, the King stipulated that each of the five parts into which his kingdom was to be divided must border on all the other four, for the King feared that if the land was divided so that one of the sons could not visit another without crossing the land of a third, the two sons with lands so separated might become estranged. In addition, the King stipulated that each pair of sons must build a road connecting their residences, and that these roads must run separately without crossing and without touching the domain of any third brother. After the King died, the five sons worked long and hard to divide the land to conform with their father's wishes, but all their efforts were in vain.

No wonder the sons failed in their task, for it can be proved that the maximum number of mutually adjacent domains on a plane (or sphere) is equal to the maximum number of points on the plane (or sphere) that can be joined by mutually nonintersecting simple curves on the plane (or sphere), and each of these maximum numbers is 4.

324° *Father and son.* Möbius was the father of a well known neurologist, whose book dealing with the "physiologically weaker mind of women" won much greater notice than any of the father's sounder mathematical works.

325° *A tea-and-sugar problem.* One summer long ago at a luncheon at the University of Chicago, Louis Mordell, the famous British number theorist, asked: "How can you distribute 14 lumps of sugar in three cups of tea, and have an odd number of lumps in each cup?" After a while he disclosed the answer: "Put one lump in the first cup, one in the second, and 12 in the third—12 lumps of sugar is an odd number of lumps to put in one cup of tea!"—I. A. BARNETT

326° *Trivial.* I once attended a seminar of Marston Morse's in which Tomkins was presenting a paper. He said, "This is trivial," and Morse interrupted coldly to say, "What are the trivial reasons?" It took half an hour to get them.—RALPH P. BOAS

327° *The relation between A and B.* Marston Morse and I were Peirce Fellows at Harvard in 1919–1920. It was customary then for the mathematics group to meet once a week. After a lecture, refreshments were served. The janitor came in with soft drinks, cheese, and so on, and remained till everyone left. One day Morse asked the janitor what he got out of the talks. His reply was: "Half the time is spent in proving $A > B$, the other half in proving $A < B$, when all the time they know $A = B$."—I. A. BARNETT

328° *Newton and the keeping of accounts.* Rev. J. Spence, in his *Anecdotes, Observations, and Characters of Books and Men* (1858) makes the following interesting comment about Sir Isaac Newton: "Sir Isaac Newton, though so deep in algebra and fluxions, could not readily make up a common account: and, when he was Master of the Mint, used to get somebody else to make up his accounts for him."

329° *Newton's epitaph.* Who, by a vigor of mind almost divine, the motions and figures of the planets, the paths of comets, and the tides of the seas first demonstrated.

330° *Wilson on monuments to Newton and Shakespeare.* John Wilson (1741–1793), while still an undergraduate at Cambridge University, discovered the pretty theorem of elementary number theory that today bears his name (if p is a prime, then $(p - 1)! + 1$ is a multiple of p). In spite of this early discovery, and the fact that he became senior wrangler in 1761, he failed to do anything further of significance in the field of mathematics. Commenting on monuments to perpetuate the memory of Newton and Shakespeare, he wrote: "A monument to Newton! a monument to Shakespeare! Look up to Heaven—look into the Human Heart. Till the planets and the passions—the affections and the fixed stars are extinguished—their names cannot die."

331° *Pope and Hill on Newton.* In an epitaph intended for Sir Isaac Newton, Alexander Pope had written:

> Nature and Nature's laws lay hid in night;
> God said, "Let Newton be!" and all was light.

Later, altering this somewhat, Aaron Hill, in his *On Sir Isaac Newton*, wrote:

> O'er Nature's laws God cast the veil of night,
> Out blaz'd a Newton's soul—and all was light.

PEANO TO SWIFT

332° *Peano and his space-filling curve.* In 1890, Giuseppi Peano (1858–1932) constructed the first continuous curve passing through every point of a square, thus showing that a continuous curve need not be one-dimensional, but can be area-filling. The curve is defined as the limit of an infinite sequence of curves, and Peano had one of the curves in the sequence put on the terrace of his home, by means of black tiles on white.

Professor Hubert C. Kennedy of Providence College in Providence, Rhode Island, says that Ugo Cassina has attested to the above story and has told just which curve of the sequence was used. He was also assured by Peano's neice, Giuseppina Peano, that she saw the figure. Professor Kennedy himself, however, was unable to locate even the house when he visited Italy a few years ago. The late Alessandro Terracini has reported in his autobiography that he too was unable to find the house.

Peano's curve has (an infinite number of) multiple points. Subsequently, continuous area-filling curves without this fault were constructed by David Hilbert (1891), Waclaw Sierpiński (1912), and others. Figures 30 (a), (b), (c) show the first three approximations p_1, p_2, p_3 of the Hilbert curve.

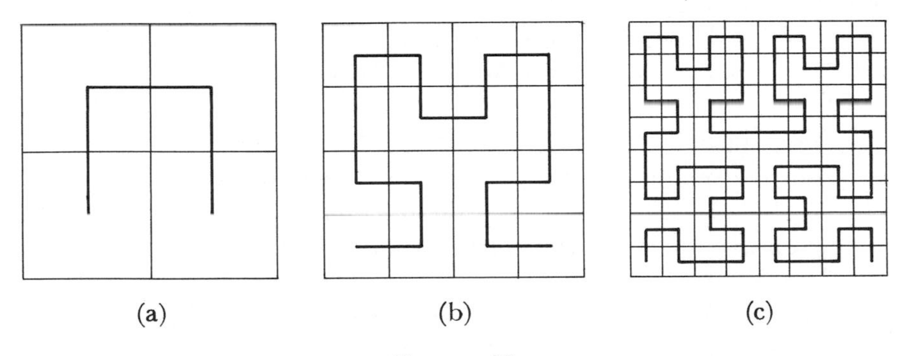

FIGURE 30

Hilbert's curve is given by p, where

$$p = \lim_{n \to \infty} p_n.$$

In imitation of Peano, the approximation curve p_3 would make a very attractive design made with tiles on a terrace or with hedges in a garden.

333° *Proper punishment.* The following episode happened when I was editing *Mathematical Reviews*. We once received a paper in Gaelic written by one R. A. MacFhraing. I looked in the file of reviewers and found only one reader of Gaelic, R. A. Rankin, so I sent the paper to him. In due course I got a review, with an attached note, "I suppose you realize whom this is really by." Then I caught on: MacFhraing is Rankin in Gaelic (and is pronounced, approximately, MacRank). I printed the review (you can look it up in the index), and wrote to Rankin that it was proper punishment for writing in Gaelic that he should have to review his own paper.—RALPH P. BOAS

334° *Point of view.* In the summer of 1945, Professor Joseph Fels Ritt was teaching a first course in function theory to Columbia University students. At the end of a blackboard demonstration he stepped back and asked, "Have I made myself clear? Are there any questions?"

"Professor Ritt," complained one student, "you stood in front of the blackboard, so I could not see what you were explaining."

"Well," remarked Professor Ritt, drily, "I try to make myself clear, but I can't make myself transparent."—ALAN WAYNE

335° *Russell's nightmare.* Bertrand Russell (1872–1970) has told of a horrible nightmare he had. In his dream he was in the great library of the British Museum, and an attendant with a large disposal bucket on wheels was slowly going up and down the aisles between the stacks. The attendant would pick a book off the shelf, judge it for a moment, and then either replace it on the shelf or drop it in the disposal bucket. At length the attendant drew from the shelf the monumental *Principia mathematica*, written by Russell and A. N. Whitehead. The writing of this work had called for a painfully exhausting effort on the part of the authors, and in his dream Russell somehow knew that this was the very last copy of the work in existence. The attendant looked at the title of the work, seemed about to drop the book into the disposal bucket, changed his mind and started to restore it to the shelf, vacillated by again withdrawing it, hesitated—and right then Russell awoke from his dream.

336° *Why Shï Huang-ti burned the books.* In 213 B.C. Emperor Shï Huang-ti decreed that all books of knowledge, including books on mathematics and related topics, be burned. Realizing that many scholars would not willingly burn their books, he further decreed that anyone caught disobeying the order would be branded and sentenced to four years of hard labor on the Great Wall. Even so, some 460 scholars banded together and defied the Emperor, and as a penalty for this defiance the Emperor had those scholars buried alive.

Now why did Emperor Shï Huang-ti engage in such wanton destruction and inhuman cruelty? It is said that he had delusions of grandeur and was determined to be remembered as the greatest of all emperors. Among other things, he wanted to become famous in history as the emperor during whose reign knowledge increased most rapidly. His strange way of attaining this end was to have all the books burned. For, he reasoned, if in the years to come there should be no books in all China that were written before his reign, but many books written during his reign, people would think that knowledge began with him.

337° *A malapropism.* On page 28 of the first edition of his excellent *A Concise History of Mathematics*, Dr. Dirk J. Struik, now Professor Emeritus of M.I.T., refers to "the volume of a frustrated pyramid." It may well be frustrating to a pyramid to have its top cut off. Nevertheless, in the third revised edition of the book, the reference is changed to "the volume of a truncated pyramid."

338° *One for the book.* Dr. Shlomo Sternberg of Harvard University tells about a graduate student of mathematics who stopped in at a bar for a glass of beer (in the neighborhood of Johns Hopkins University) and put his book, a treatise in advanced mathematics, on the top of the bar while he was drinking his beer. Another patron of the place, seeing the word "mathematics" in the title, picked up the book and glanced through it. "Oh, ho!" he exclaimed, "how much I've forgotten!"—ALAN WAYNE

339° *Sturm and his theorem.* Jacques Charles François Sturm (1803–1855) was a native of Geneva, Switzerland, and he succeeded Poisson in the chair of mechanics at the Sorbonne. In 1829 he published, without a proof, his celebrated theorem concerning the number of real roots, lying within a given number interval, possessed by a given polynomial equation having real coefficients. This discovery by Sturm completed the theoretical solution of a long-standing difficulty in the theory of equations that had called forth the energies of a galaxy of great mathematicians since the time of Descartes. Proofs of the theorem were furnished in 1830 (by Andreas von Ettinghausen), 1832 (by Charles Choquet and Mathias Mayer), and 1835 (by Sturm himself). The discovery cost Sturm considerable effort, and he was justly proud of his achievement. When, in his lectures, he came to the presentation of the theorem, he would say, "Here is the theorem whose name I bear."

340° *The astonishing prediction in "Gulliver's Travels".* [The following is adapted, with permission, from the article, by Howard Eves, of the same title that appeared in the Historically Speaking section of *The Mathematics Teacher*, December, 1961, pp. 625–626.]

In *A Voyage to Laputa,* Dean Swift has Gulliver report on some of the remarkable achievements of the Laputians. Gulliver says, in part: "[The Laputian astronomers] spend the greatest part of their time in observing the celestial bodies, which they do by the assistance of glasses far excelling ours in goodness.... This advantage hath enabled them to extend their discoveries much further than our astronomers in Europe; for they have made a catalogue of ten thousand fixed stars, whereas the largest of ours do not contain above one third of that number. They have likewise discovered two lesser stars, or satellites, which revolve about Mars, whereof the innermost is distant from the center of the primary planet exactly three of his diameters, and the outermost five; the former revolves in the space of ten hours, and the latter in twenty-one and an half; so that the squares of their periodic times are very near in the same proportion with the cubes of their distances from the center of Mars, which evidently shows them to be governed by the same law of gravitation, that influences the other heavenly bodies."

Before going on, let it be noted that *Gulliver's Travels* was originally published in 1726.

In 1877, a good 150 years after the publication of *Gulliver's Travels,* the American astronomer Asaph Hall, using the finest telescope of the time, the 26-inch refractor at the Naval Observatory in Washington,

D.C., discovered that Mars possesses two small satellites or moons. These satellites, which have diameters probably under ten miles, are so near the planet Mars that they can be seen only with sufficiently large telescopes, and even then only at favorable times. Suitable telescopes for this observation were not made until about a century after the publication of *Gulliver's Travels*.

Asaph Hall named the two satellites Phobos and Deimos. Phobos, which is the inner satellite, completes a sidereal revolution in 7 hours and 39 minutes, and revolves about Mars at a distance of about 5800 miles from the center, or about 3700 miles from the surface, of the planet. The sidereal period of Deimos is 30 hours and 18 minutes and revolves at about 14,600 miles from the center of Mars. Deimos is smaller than Phobos and is only about one third as bright.

Of the above surprising coincidences, Charles P. Olivier (former director of the Flower and Cook Astronomical Observatories at the University of Pennsylvania) has written: "When it is noted how very close Swift came to the truth, not only in merely predicting two small moons, but also the salient features of their orbits, there seems little doubt that this is the most astonishing 'prophecy' of the past thousand years as to whose full authenticity there is no shadow of doubt.... [Phobos's] period of less than eight hours obliges it to rise in the west and set in the east. In this...respect it is unique among all bodies in the universe, so far discovered. Yet Swift had this fact also included."

The remarkableness of it all does not quite end here, for these same two satellites were also mentioned by Voltaire (1694–1778) in his story of *Micromegas*!

SYLVESTER TO WHITEHEAD

341° *An incident in Sylvester's early life.* The most distinguished of the private pupils who studied mathematics under Sylvester in the hard days of his early life was a young woman named Florence Nightingale, later to become world famous as a reformer of hospital nursing. Sylvester was in his late thirties and Miss Nightingale was six years younger. Sylvester was able to give up private tutoring as a means of earning a living in 1854, the same year that Miss Nightingale left England for the battlefield of the Crimean War.

342° *An incident in Sylvester's later life.* Knowing very little about elliptic functions, and wishing to apply them to a certain part of the theory of numbers that was currently interesting him, Sylvester once, in later life, engaged a young man to instruct him on the subject. After only a very few lessons Sylvester abandoned the attempt and instead began lecturing to the young instructor on his own latest algebraic discoveries.

343° *The mathematical Adam.* In volume 37 (1887–1888), page 162, of *Nature*, Sylvester quite truthfully wrote: "Perhaps I may without immodesty lay claim to the appellation of mathematical Adam, as I believe that I have given more names (passed into general circulation) of the creatures of the mathematical reason than all the other mathematicians of the age combined."

Perhaps Professor Edward Kasner can be called the mathematical Adam of more recent times.

344° *Sylvester's papers.* Sylvester finished his mathematical papers with difficulty. His handwriting was very poor and a great trouble to the printer. A paper submitted to a journal would be scarcely out of his hands when he would forward to the printer a sheaf of alterations, corrections, and additions. For several posts thereafter, further directions would be transmitted to the harried editor and printer.

345° *Sylvester as a teacher.* Sylvester was a most unmethodical teacher. He announced: "Three lectures will be delivered on a New Universal Algebra." At the conclusion of the third lecture, he said that he must extend the sequence to twelve lectures. In the end, the lectures on a New Universal Algebra occupied the rest of the academic year.

The following year he was to offer a course in substitution theory, using Netto's treatise on the subject. All the students bought the text, and for three lectures Sylvester faithfully followed the book. Then he became interested again in some matrix problems and announced that he must give over one lecture a week to matrices. The one hour proved insufficient, and after two more weeks Netto was forgotten and never mentioned again.

Once he came in and said: "Here is a truly remarkable theorem.

I haven't proved it yet, but I feel absolutely certain that it is true. Here are some of its consequences." At the next lecture he announced that the remarkable theorem of last time turned out to be false, and offered a splendid corrected replacement for it, from which some wonderful discoveries followed.

The students sat there, watching mathematics being created, hot from the forge.

346° *The mathematics of golf.* Tait had an interest in golf and was the first to study the flight of a golf ball mathematically. His explanation of the secret of long driving was scoffed at at first. He showed that spin is an important factor. "In topping, the upper part of the ball is made to move forward faster than does the center, consequently the front of the ball descends in virtue of the rotation, and the ball itself skews in that direction. When a ball is undercut it gets the opposite spin to the last, and, in consequence, it tends to deviate upwards instead of downwards. The upward tendency often makes the path of a ball (for a part of its course) concave upwards in spite of the effects of gravity." It is the underspin that prolongs both the range and the time of flight. Without spin a ball cannot combat gravity to any great extent, but with spin it can travel remarkable distances.

Tait was fond of golf, whereas Hermann von Helmholtz, who visited Tait in Scotland, "could see no fun in the leetle hole." One of Tait's sons became a brilliant golfer.

347° *William Thomson and the old Highlander.* S. P. Thomson, in his *Life of Lord Kelvin* (1910), narrates an entertaining story that was told by Lord Kelvin himself when dining at Trinity Hall.

It seems that a Highland country lad attended the university and had done very well, receiving, at the close of the session, prizes in both mathematics and metaphysics. The lad's old father left the farm and came up to the college to see his son awarded the prizes. Professor William Thomson, later to become Sir William Thomson and Lord Kelvin, showed the old man around the college. "Weel, Mr. Thomson," asked the old farmer, "and what may these mathematics be for which my son has getten a prize?" Professor Thomson told him that mathematics was calculating and reckoning with figures. "Oo ay," said the old man, "he'll ha' getten that fra' me; I were ever a braw hand at the countin'." Then he asked, "And what, Mr. Thomson, might these metapheesics be?" Professor Thomson tried to explain how metaphysics attempts to express the indefinite. "Oo ay," said the old farmer, "maybe he'll ha' getten that fra' his mither; she were aye a bletherin' body."

348° *William Thomson and his students.* Unable to meet his classes one day, Lord Kelvin posted the following notice on his lecture-room door:

"Professor Thomson will not meet his classes today."

The arriving students decided to play a prank on the professor, and erasing the "c" they left the notice to read:

"Professor Thomson will not meet his lasses today."

Next day, the assembling students, anticipating some fun from their prank, were chagrined to find the professor had outsmarted them. The note now read:

"Professor Thomson will not meet his asses today."

349° *Titles.* Peter Guthrie Tait dubbed James Clerk Maxwell dp/dt, for in thermodynamics $dp/dt = JCM$, where C denotes Carnot's function. Maxwell, on his part, denoted Sir William Thomson (Lord Kelvin) by T and Tait by T', so that Thomson and Tait's *Treatise on Natural Philosophy* became referred to as *T and T′*.

350° *What is a mathematician?* Perhaps the most frequently repeated anecdote about Lord Kelvin concerns the occasion when one day he asked his class if they knew what a mathematician is. Stepping to the blackboard he wrote

$$\int_{-\infty}^{\infty} e^{-x^2}\,dx = \sqrt{\pi}.$$

Turning to the class and pointing to what he had written, he said: "A mathematician is one to whom *that* is as obvious as twice two makes four is to you."

351° *Von Karman's absent-mindedness.* [The following item is adapted, with permission, from George Pólya's article "Some mathematicians I have known," which appeared in the August-September 1969 issue of *The American Mathematical Monthly*, pp. 746–753.]

I heard the following from Theodore von Karman himself. Still, I would not swear that it actually happened; he liked good stories too much, and the best stories do not happen, they are invented. At that time he had a double position: He was professor at Aachen in Germany and also lectured at Cal Tech in Pasadena. As an important aeronautical engineer, he was consultant to several airlines, and so he got free transportation whenever he found an unoccupied seat on a plane of one of these lines. So he commuted more or less regularly between Aachen and Pasadena. He gave similar lectures at both places. Once he was somewhat tired when he arrived in Pasadena, but started lecturing. That was not so difficult: He had his notes which he also used in Aachen. He talked, but as he looked around he had the impression that the faces in the audience looked even more blank than usual. And then he caught himself: He was speaking in German! He became quite upset. "You should have told me—why did you not tell me?" The students were silent, but finally one spoke up: "Don't get upset, Professor. You may speak German, you may speak English, we will understand just as much."

352° *A late comer.* It is noteworthy that Karl Wcicrstrass, unquestionably one of the world's greatest mathematicians, spent a large part of his life teaching in school, and did not start his career as a university professor until his forty-ninth year, an age by which many mathematicians have already ceased to be creative. He never regretted his schoolroom experience, and he has become known in the history of mathematics as the greatest teacher among the top-flight research men of the subject.

353° *Late for class.* One morning Karl Weierstrass failed to appear to teach his eight o'clock class at the Braunsberger gymnasium, and the waiting students were making considerable noise. The director of the gymnasium went in person to Weierstrass's dwelling to ascertain the cause of the teacher's absence. On knocking he was invited to come in. There he found Weierstrass working by a dim lamp in a darkened room and totally unaware that it was daylight outdoors. Weierstrass, who had completely lost track of time, had worked the night through. When the director told him of the noisy students waiting for him, he asked the director to excuse him, as he was on the verge of an important mathematical discovery that he hoped would surprise the scientific world.

354° *Stamp of approval.* At Oxford University, when a professor concludes a course, it is a custom for the students to pound the floor with their feet as a "tribute" to the teacher for his fine teaching. On one occasion, when A. N. Whitehead had finished his last lecture, the pounding of the feet was so enthusiastic that in the room below, where a professor of logic was lecturing, the ceiling began to fall. The professor of logic remarked: "I am afraid that the premises will not support Dr. Whitehead's conclusion!"

One is reminded of the two washerwomen who frequently leaned out of their windows on opposite sides of an alley and quarreled with one another. You see, they could never agree because they were arguing from different premises.

NORBERT WIENER

NORBERT Wiener (1894–1964) was for many years the adored exhibition piece of the M.I.T. campus. A prodigy, he received his A.B. degree from Tufts College at the age of fourteen, and his Ph.D. from Harvard when he was eighteen. His first teaching post was at the University of Maine in Orono, where he spent an unhappy year wrestling with the "lumberjack minds" of his students. He next worked for the General Electric Company at Lynn, Massachusetts, then for the Encyclopedia Americana in Albany, from where he was called to the Aberdeen Proving Grounds to serve as a computer during the war. After the war he accepted a post on the old Boston *Herald* as a feature writer. Then, in 1919, he returned to academic life by joining the faculty at M.I.T., there to remain, except for several leaves of absence, for the next forty years, and to become the Institute's favorite tradition. He was a short, bearded, stocky man, with quick movements, and always avidly reading from a book in his hands as he walked along. In time his brilliance and his eccentricity became woven into the M.I.T. campus mythology, and a host of stories and legends sprang up about him. Unbelievable as many of these stories are, almost every one is sworn to by some "on the spot" observer. Professor Wiener was a voluminous and fluent writer, a gifted linguist, an entertaining lecturer, an ardent mountain climber, very empathetic, and the possessor of a highly creative mind.

355° *Norbert Wiener parks his car.* One day Norbert Wiener parked his car in a large lot with many other vehicles, and went inside the adjoining building to attend a conference. Upon the conclusion of said conference, Wiener came outside to drive off, but was unable to locate his car, in fact was not even sure of what it looked like. So he waited patiently until all the cars had been driven off but one—and that was his car.—JOHN K. MOULTON

356° *Norbert Wiener moves.* It happened that the Wieners moved to a new house, in the same neighborhood as the former residence. Knowing her husband's absent-mindedness, Mrs. Wiener gave him careful and written instructions for reaching the new house. However at the close of the day, Professor Wiener could not find the written instructions and of course did not remember them. Hence, seeking something familiar, he set off in the direction of his former residence. Presently he spied a young child and asked her: "Little girl, can you tell me where the Wieners have moved to?" "Yes, Daddy," came the reply, "Mommy said you'd probably be here so she sent me over to show you the way home."—JOHN K. MOULTON

357° *Wiener finds the sought word.* There is a widely told story about Norbert Wiener working at a table in the campus lounge. He was observed studying a paper in front of him on the table with tremendous concentration. He would every now and then get up from the table, pace a bit, and then return to the paper. Everyone was deeply impressed by the enormous mental effort that was reflected on Wiener's face. Once again he rose from his paper, took a few rapid steps about the room and collided with a student. The student said, "Good afternoon, Professor Wiener." Wiener stopped, stared, clapped a hand on his forehead and said, "Wiener—that's the word," and ran back triumphantly to the table to fill the word "wiener" in a crossword puzzle that he was working.

358° *Norbie.* George Pólya once overheard a conversation between Norbert Wiener and a friend. "Confess," said Wiener, "that you call me Wienie behind my back." "No," said the friend, "we call you Norbie."

359° *Wiener's start in mathematics.* It seems that it was an unsatisfactory grammar school report card that started Norbert Wiener on his mathematical career. When Norbert's father, who was a professor of Slavic languages at Harvard University, saw his son's low grade in arithmetic, he took the boy in hand for some concentrated tutoring in mathematics. The boy found the subject was very easy. "It was just laziness from that point on," Norbert later said, "for I knew it would take less work to specialize in mathematics than in any other subject."

360° *Professor Wiener's famous letter.* In 1947, Professor Wiener wrote a letter to a research scientist of one of our country's large aircraft corporations. The scientist had inquired about Professor Wiener's war research, which had been rather extensive. The letter, entitled *A Scientist Rebels*, was published in the *Atlantic Monthly* and subsequently in a number of anthologies, and it stated many of Professor Wiener's acquired views on the place of science and the scientist in a troubled world. The letter in part reads:

> The policy of the government . . . during and after the war, say in the bombing of Hiroshima and Nagasaki, has made it clear that to provide scientific information is not necessarily an innocent act, and may entail the gravest consequences. One, therefore, cannot escape reconsidering the established custom of the scientist to give information to every person who may enquire of him. The interchange of ideas, which is one of the great traditions of science, must of course receive certain limitations when the scientist becomes an arbiter of life and death.

Dr. Wiener added that he would no longer publish work that

might do danger in the hands of "irresponsible militarists." And several months later, in the *New York Herald-Tribune*, he buttressed his position by declaring he would no longer engage in government research. He decried the use of science as a weapon, and he asserted that he would not work on any project that might mean the ultimate death of innocent people.

One is reminded of Albert Einstein when, in a discussion following the development of the H-bomb, the grim matter arose of the very possible destruction of the world should opposing political ideologies resort to atomic force. With an expression of utter horror on his face, in a very sad voice he said, "No one would ever again hear the music of Mozart."

INDEX

References are to items, *not* to pages. A number followed by the letter *p* refers to the introductory material just preceding the item of the given number (thus 227*p* refers to the introductory material immediately preceding Item 227°).

INDEX

INDEX

Index

Index

Index

INDEX

Index